PERCEIVING
TIME

PERCEIVING TIME

A Psychological Investigation
with Men and Women

THOMAS J. COTTLE

A WILEY-INTERSCIENCE PUBLICATION

JOHN WILEY & SONS, New York • London • Sydney • Toronto

Library of Congress Cataloging in Publication Data:

Cottle, Thomas J.
  Perceiving time.

  "A Wiley-Interscience publication."
  Bibliography: p.
  Includes index.
  1. Time perception. 2. Personality. 3. Sex
difference (Psychology) I. Title.
BF468.C58   153.7'53   76-18768
ISBN 0-471-17530-7

Printed in the United States of America

10 9 8 7 6 5 4 3 2 1

FOR KAY M. COTTLE

PREFACE

A little more than a decade ago J. B. Priestly wrote an elegant treatise on the meaning of time and what is known about our perceptions of it.[1] Clearly the topic troubled Priestly, for early in his book he concluded that one cannot resolve the various meanings, experiences, and theories of time. The problem of trying to understand time is too complex. The scholastics had wondered about temporality and causality. Augustine, trying to understand the meaning of time, confided that when he tried to articulate his thoughts on it, he was left with nothing.[2]

Short of confessing that his entire effort was misguided, Priestly attempted to summarize the enormous problem one faces in studying the meaning of time.

> For time as we know it, we need both change and not change, some things moving and others apparently keeping still, the stream flowing and its banks motionless. This may seem too obvious a point to make. . . . One philosopher tells me that all is flux, nothing remaining the same. But how can he know this? If everything is changing, including himself, how can he know that anything is changing? There could be no standard of comparison, no point of reference.[3]

Although few can disagree with Augustine, Emerson, and Priestly that time is a phenomenon almost impossible to explore, the literature dealing with the meaning of time is filled with valuable theories and discoveries. Philosophers, physicists, and social scientists have made significant contributions to an understanding of what time means to us. In some cases these contributions may seem to be esoteric and not particularly helpful in understanding the meaning of time in our own lives. In other cases we may feel that a particular researcher was so involved in the *method* of discovery that the discovery itself is insignificant by comparison.[4] Still, the body of information on the meaning of time is enormous, and while no one would say with certainty that he or she has truly understood the meaning of time, a certain amount is known.[5] We hope that the research presented in the following chapters will help in the understanding.

Our focus in this book is on people's perceptions of time. More precisely, we seek to know how people perceive and feel about the five major periods of time that comprise the temporal horizon—the personal past, or the time from one's birth to the present; the present; the personal future, or the time from the present to the moment of one's death; the historical past, or the time prior to one's birth; and the historical future, or the time following one's death. In considering the temporal horizon, we wish to know, among other things, what are the chronological units that define the various time periods? What is the relationship of one time zone to another? What portion or portions of the temporal horizon preoccupy people's attentions most? What sorts of fantasies about the past, present, and future do people have, such as the wish to recover time from the past or to know the future?

We want to know something about the source of these perceptions. Do they derive in some way from one's culture, from the values one learns and respects, for example, the value to achieve? Do they derive from the definition of one's social role or, more specifically, from the meaning of being a man or a woman? Indeed, one question we address in every chapter is whether characteristically masculine and feminine perceptions of time exist. Do perceptions of time derive from the qualities of one's personality? Does anxiety or measured intelligence, for example, affect one's perceptions of the temporal horizon?[6]

We may also turn these questions around and ask: Might one's perceptions of time affect one's intelligence, the state of one's anxiety, one's sense of one's masculinity or femininity, or one's values? To ask the questions in this way is to go against the tradition of psychological studies of time perception. It is to suggest that anxiety, intelligence, or the value to achieve derive from perceptions of time. The implication of this line of reasoning is that one's involvement with the temporal horizon—one's experience of living in and through time—contributes significantly to the development of one's anxiety, one's masculinity or femininity, and the degree to which one values achievement.

These are complex questions, but little about the meaning of time is simple. Perceptions of only one time zone, the future, for example, may become so complicated that the perceptions held by one person or one group of people might appear totally unrelated to perceptions held by another person or group. One of our major tasks in this work is to describe the patterns of perceptions of time so that seemingly unrelated perceptions may be clarified and understood as a totality.

The necessity for clarification, however, should not be interpreted to mean that as a topic of inquiry the perception of time remains unexplained. Considerations of time are found in the work of both ancient and contemporary philosophies. They appear in literally hundreds of studies performed by physicists, biologists, and psychologists as well as in the research of sociologists and anthropologists. But few social scientists have attempted to make systematic assessments of even the chronological meanings of the past, present, and future. One can legitimately say, for example, that a man is past-oriented if he answers a question in a way that previously has been designated to mean past-oriented. Indeed, as anthropologists have shown, nothing is wrong with distinguishing between cultures on the basis of how they conceive of the past, present, and future. Yet, how do we know where on the horizon of time people divide the past from the present and the present from the future? Some people, after all, might feel that the present turns into the future 10 years from now, thereby indicating that the period of the present will last for about a decade. Other people might say that the future lies at the bottom of this page, for anything that has not yet been experienced is in the future. Because of variations in perceptions, in our study we have tried to examine the basic definitions of time and the time zones we call the past, present, and future.

The 11 chapters of the book represent a continuing story of our research into men's and women's perceptions and evaluations of the temporal horizon; each chapter is organized around one or two major perceptions and thus provides a separate story. An inquiry, method for researching a particular perception of time, and a discussion of the findings are presented in each chapter. In Chapter 11, we look at our study in its entirety and offer some speculations and conclusions.

One final word before we begin: The problems of doing research on perceptions of time become particularly critical because so few phenomena are as important to an understanding of the meaning of human life. Keeping in mind the difficulties and discouragement expressed by Augustine, Emerson, and Priestly, we must always maintain a modest approach to a study of time. Only a bit of knowledge is gained in even the most momentous study; only a small part of the mystery of time is revealed in even the most scientific inquiry. To those men and women who consented to participate in this research, a special thank you and a hope that they benefitted in some way from their efforts and generosity.

THOMAS J. COTTLE

London, England
June 1976

NOTES

1. J. B. Priestly, *Man and Time* (Garden City, N.Y.: Doubleday, 1964).
2. Saint Augustine, *Confessions*. Tr. by J. F. Pilkington (New York: Dial, 1964).
3. Priestley, *op. cit.* p. 64.
4. In this context, see Alfred North Whitehead, *Modes of Thought* (Glencoe: The Free Press, 1968).
5. See Leonard W. Doob, *Patterning of Time* (New Haven: Yale University Press, 1971); and R. E. Ornstein, *On the Experience of Time* (Baltimore: Penguin, 1970).
6. On this point see J. E. Orme, *Time, Experience and Behavior* (New York: American Elsevier, 1969); J. W. Dunne, *This Serial Universe* (London: Macmillan, 1938); and G. J. Whitrow, *What is Time* (London: Thames and Hudson, 1972).

ACKNOWLEDGMENTS

I would like to thank the Department of Sociology, The University of Chicago, and especially Professors Fred L. Strodtbeck, Jacob W. Getzels, Donald Levine, and John M. Butler; The Children's Defense Fund of the Washington Research Project, and especially Marian Wright Edelman and Sally Makacynas; also Valda Aldzeris, Abby Bryony, Robert F. Bales, Robert J. Beck, William Bezdek, Anthony Costonis, Carl N. Edwards, Karla Heuer, Peter Howard, Jerome Kagan, John F. Marsh, Jr., David C. McClelland, Stanley Milgram, Gerald M. Platt, William Morgan, John Muller, Jack Sawyer, and Cynthia Sparrow.

Special thanks must go to James S. Dalsimer, Stephen L. Klineberg, Joseph Pleck, who contributed especially to Chapter 8, and David Ricks. All of these people have helped me in a great many ways. My editors Eric Valentine and Judith Cuddihy have worked indefatigably on the book.

Kay M. Cottle, my wife, as always my colleague, made certain that the work was not only completed, but done properly and ethically.

Finally, I wish to thank the corpsmen and corpswomen of the Great Lakes Naval Training Center as well as their commanding officers. I shall always be indebted to them.

T. J. C.

CONTENTS

1. Introduction 1

2. Linear and Spatial Conceptions of Time, 6

3. Orientations to the Past, Present, and Future: The Experiential Inventory, 20

4. The Perceived Duration of Time Zones: The Duration Inventory, 34

5. Fantasies of Temporal Recovery and Knowledge of the Future: The Money Game, 50

6. Future Orientations and Avoidance: The Affect of Social Roles on Perceptions of Time, 66

7. A Spatial Conception of Time: The Circles Test, 85

8. A Linear Conception of Time: The Lines Test, 102

9. Duration, Potency, and Relatedness: Examining Perceptions of the Present, 131

10. The Affect of Valuing Achievement and Manifest Anxiety on Perceptions of Time, 149

11. Contrasting Men's and Women's Perceptions of Time: Conclusions and Speculations, 169

Bibliography, 195

Tables, 215

Index, 261

PERCEIVING TIME

CHAPTER ONE

INTRODUCTION

The investigations upon which this book is based explore the ways people perceive the temporal horizon. We are inquiring into fundamental perceptions of time and attitudes toward it, the periods of time that constitute the phases of a person's life, and the historical past and future. Our goal is to understand how linear and spatial conceptions of time and personality variables, social roles, and cultural values influence our perceptions of time. In addition, we wish to know how, in objective and subjective terms, people perceive these aspects of time and what significance they attribute to them.

In each of the following chapters we examine one aspect of our study. The final chapter contains our speculations and conclusions.

In Chapter 2 we briefly review the two essential conceptions of time considered in this book—the linear and spatial. We see how external stimuli, such as light and sound, the anticipation of danger, and the experience of success and failure in the performance of a task affect perceptions of time. In our review of temporal conceptions we examine the significance of immediate experience, the recollection of past experience, the expectation of future experiences, and how these three types of experience might be related.

We continue our examination of the degree to which we accord significance to the past, present, and future in Chapter 3. We recall that in any view of time we must include a consideration of memory of the past, the experience of present sensations, and our expectations of the future. Philosophers and psychologists disagree over which of these three major time zones is the most important for understanding our perceptions of time,[1] raising the additional question of whether we may accurately categorize people in terms of their orientations to the past, present, or future. The empirical part of our study begins in Chapter 3 with an investigation of the degree to which people orient themselves to the past, present, and future. Using an instrument called the Experiential Inventory, we attempt to group people on the basis of where in time they locate the important experiences of their lives.

In Chapter 4 we continue to discuss variations in estimations of time-zone duration, and we examine empirically the units of time that people use in determining the duration of the distant and near past, the present, and the distant and near future. To make this examination we developed an instrument called the Duration Inventory, which is described in detail in Chapter 4.

Fantasy is another approach to the phenomenon of time. While not actually indicative of the way time is represented to us in experience, it is one we often take. Experimental conditions influence our estimations of time passage, but this does not imply the fantasy of changing the speed of time. In Chapter 5 the fantasy we discuss is the notion of being able to recover past time and to know the future before it arrives. The instrument used in our inquiry, the Money Game, allows us to discover the degree to which we like to indulge in this form of fantasy.

Chapter 6 is devoted largely to a study of perceptions of, and attitudes toward, the future. Among the topics we examine are the type of preparations we make for the future, our willingness to make predictions about it (for which we employ an instrument called the Future Commitment Scale), the perceived "power" of the future, and how our various perceptions of the future relate to each other. We begin to explore the affect of social roles on temporal perceptions based on a series of findings that men and women differ in the way they perceive time. We base our explanation of these differences on a definition of the types of roles men and women in our society are obliged to play.

In Chapters 7 and 8 we explore in greater detail the issue of linear and spatial conceptions of time. In Chapter 7 we focus on spatial conceptions and, more specifically, on the perceived relationship—what Bergson called the correlation—between the past, present, and future. The instrument presented in Chapter 7, the Circles Test, allows us to examine spatial aspects of time and affords us another opportunity to examine the varying significance that people attribute to the past, present, and future.

In Chapter 8 we turn to the linear conception of time. With the use of an instrument called the Lines Test, we examine topics such as the linear duration of the past, present, and future and the distinction people make between the time of their own lives and the time of the historical past and the historical future. In our effort to understand the sources of perceptions of time, we also examine the degree to which age, sex, and social class affect the way we perceive linear durations of time zones.

The period called the present is of utmost significance if we are to understand how human beings experience time. Indeed, some writers have argued that the only genuine time zone is the present because we exist only in the present instant—the here and now. There is so much disagreement on this point that we devote all of Chapter 9 to an examination of perceptions of the present.

Chapter 10 is a continuation of the study of the sources of time perceptions and the external factors that influence perceptions of time. Throughout the book we discuss the effect of social role on time perceptions, but in Chapter 10 our attention turns to the influence of cultural values. We are concerned with the valuing of achievement and with psychological factors such as measured intelligence and manifest anxiety on how people perceive the temporal horizon.

Chapter 11 is devoted to speculations and conclusions.

POPULATION AND PROCEDURE

Before turning to a discussion of what the past, present, and future means to people and to the question of whether people can accurately be categorized according to their preoccupation with one of these time zones, let us describe the sample of men and women whose perceptions and attitudes constitute this study, and our rationales for selecting this particular sample of people.

Because our study focuses on perceptions of time and the effect on these perceptions of variables such as values and social roles, we made it a point to hold constant variables such as social class, occupation, and level of education. We therefore selected a sample of people of similar social background and level of education engaged in a similar occupation. We also faced the problem of finding people who would have the time to respond to the sorts of instruments and questionnaires we had developed. We decided to use United States Navy personnel stationed in a large military complex outside a midwestern city.

These people were engaged in a 16-week voluntary program of medical training consisting of elementary medical courses and a special patient-care program, much like an abbreviated course of instruction for nurses. While many of the corpsmen and corpswomen went on to college after completion of their military tenure, few went on to medical school. As a group, these people provided the sociological homogeneity we needed for our sample. We examined their navy records and found that all of the students had completed high school[2] and came from comparable middle-class backgrounds. Only students between the ages of 17 and 21 were included in the sample. The students also represented almost all 50 states. Less than 1% were foreign born, and most came from either small towns or medium-sized cities. Over 50% of the sample reported that both their parents held part-time or full-time jobs, and less than 2% of the sample were married.

Testing was done in groups, each group consisting of a naval company, normally 40 to 50 persons. Testing took place during class periods.[3] Tests were administered by non-navy personnel, but students were advised that they were not required to take them. Students were assured that the inventories were not measures of intelligence or mental health and that there were no right and wrong answers; we wanted only their personal feelings and attitudes. Finally, we assured the students that their responses would be held in confidence and would not appear in their navy records.

Testing took place in the first, second, and fourteenth weeks of the medical-training program. All time-perception measures were administered in the first and second weeks. The order of tests is the order in which they are presented in the following chapters.[4] Achievement and anxiety measures were administered in the 14th week; to check reliability, the time measures were also repeated in the 14th week.[5]

Because one's outside activities may influence one's perceptions of the past, present, and future in a study of the perception of time, it may be said that navy duties and obligations were similar for all the corpsmen and corpswomen. All the students progressed along the same path toward the same goal: graduation from the program. Unlike most college students who are free to choose courses and extracurricular activities, the choices allowed navy personnel were proscribed. Thus, if the medical program affected students' concepts and perceptions of time, as we assume it must have, it should have affected all students equally.

Two final points about our population and procedures: Some might contend that the navy sample is a poor one because military activity generally is masculine in nature and hence that we selected a masculine type of woman. In response, we can say that there is no evidence that Navy women are more masculine than civilian women. Thus, if differences in perceptions between men and women occur within the sample, we may feel confident that they are not caused by the navy program. Instead, they represent perceptions of time the men and women developed well before they entered the navy. Or, we may argue that if the navy program does influence perceptions of time, the differences in perceptions that emerge between the sexes indicate that men and women do not respond identically to the program.

Our overriding concern in conducting our research was that people be able to respond to the questions we asked them. Perhaps many of the ideas presented in the following chapters seemed irrelevant to the lives of the corpsmen and corpswomen. When we asked them to draw the spatial relationships between the past, present, and future, we may well have asked them to think about issues that had no conscious meaning for them. The important point, however, was not to ask them questions or to perform tasks that made no sense whatever. As we are about to see, perceptions of time are difficult enough to understand, without a researcher confusing his subjects with seeming irrelevancies.

NOTES

1. On this point see Erik H. Erikson, *Identity, Youth and Crisis* (New York: Norton, 1968).
2. In addition, 2% had attended some college.
3. A small dropout rate existed at this particular institution, but very few people were lost from the study because of it. Flunking, for the most part, meant going back into the company behind one's original company rather than leaving the school.
4. Students were instructed to work on each inventory in the order that it occurred in the test booklet.
5. The reader will see that the actual number of people varies in some of the calculations. This is because at each testing session a small number of students were absent and only those students completing all the questionnaires and inventories were included in the final calculations.

   At the end of each testing session debriefing discussions were held in which the corpsmen and corpswomen could learn about the study. Because new companies enter the program each week, we wondered whether these incoming students would learn about the study from the companies in front of them. From the debriefing discussions, we learned that they did not.

CHAPTER TWO

LINEAR AND SPATIAL
CONCEPTIONS OF TIME

The student of time faces a difficult task. How does one even begin to carve out an area of research when the concept of time is so rich and complex?[1] When one speaks of time, is one speaking of history? Is time meant to imply aging? Is time memory? Is it the ability to retrieve something from the past while knowing that if it once existed it is now only a reflection and thus somehow changed? Does time refer to notions of succession and order, the realization that *a* comes before *b*? Or is time simulteneity, the idea of experiences or events occurring at the same "time?" Or does time refer to the study of rhythm[2]—biological rhythm as in pulse, heart or respiratory rates; sociological rhythms as in work-sleep cycles, or even the cultural rhythms of recurring ceremonies or events in the history of a people?[3] Almost any word or phrase used to speak of time has so many connotations and so many potential research implications that at some point researchers must simply select those few areas on which they will concentrate and reconcile themselves to the fact that only a small area can be covered.[4]

As a subject of research, time has been useful to students of the behavioral sciences in different ways. It continues to be treated as a variable against which, or with which, behavior may be observed. Consider, for example, a psychologist performing a study in which a person's blood pressure is taken at a particular point in time. The person is then startled or frightened. Minutes later, at a second point in time, the person's blood pressure is taken again. On the basis of such an investigation, a psychologist may report that *after* being frightened, the person's blood pressure differed from what it was *before*. In this experiment, one is looking for a change to occur over time.

Another example, also from the work of psychologists, represents what is called a psychophysical estimation study. A person is seated at a table in a dark room. Before him is a light bulb and a switch that turns the light bulb on and off. He is instructed to turn on the light bulb and leave it on for what he perceives to be three minutes. At that point he is to turn the light off.

The point of the study is to see how closely a person can come to estimating the actual passage of time. Naturally, the conditions of the experiment can be altered. All sorts of stimuli may be introduced during that period, such as noise or flashing lights; or the person might be asked to carry out some task, such as sorting pictures into various piles. The essential purpose of the experiment, however, is to determine how closely a person can estimate the actual chronological passage of time.

A second way in which the variable time has proved useful to social scientists is in the way it gives meaning to both private experience and public behavior.[5] Let us cite only a few of the many examples. According to Robin George Collingwood, during their psychological development children confront problems of nature first and problems of history later.[6] Still later children must combine the understanding of nature and history to appreciate the meaning of causality.

We can take Collingwood's notion about the learning of history one step further and provide another example of this second usage of time. Consider the importance of history in psychological reasoning, particularly in the development of psychoanalytic theory and practice. One of the important features of psychoanalysis (and of schools of psychology that emerged after the psychoanalytic school) is its reliance on the patient's recollections of his or her past to reveal the patient's present personality. Using the patient's descriptions of the past, the psychoanalyst develops a picture of the patient's present situation or problems.[7] Of course, this is a usage of time different from asking people how many minutes have elapsed since a light went on.

We might suggest that these two usages of time in the behavioral sciences rely upon the objective and subjective features of time respectively. In the psychological laboratory study one is seeking to estimate the passage of chronological time. In the psychoanalytic process one is seeking to understand the patient's recollections and his or her perceptions of experiences in time. This extremely important distinction between objective and subjective perceptions of time is a central concern throughout this book. In leading our lives we are constantly making both objective and subjective evaluations of time. We wonder whether we will be late for an appointment or for work and carefully make plans to be on time. We also wonder what the future will be like or how the past has affected our present circumstances. We are constantly aware of our objective and subjective perceptions of time.

In our work on perceptions of time we have focused on what we have called the temporal horizon—time as it moves through history to the moment of birth; to the time zones we call the personal past, present, and future; and finally to the period of time that follows our death. We want to know how people perceive the temporal horizon, how they define the major periods of time constituting the horizon, and how they feel about the temporal horizon. To a certain degree, therefore, we are following J. J. C. Smart's suggestion that students of time turn their attention to what he called the "emotional significance" that time holds for people.[8] In more general terms, the issue of studying time suggests a challenging question: Is there really a phenomenon of time that exists apart from any individual, or does the concept of time reside only in one's perceptions of it? Does time possess objective and subjective features, or do human beings bring what they feel to be objective and subjective evaluations to their understanding of time? Aware of these complex questions, Frederich Kummel wrote:

The problem concerning the nature of time cannot be separated from the many forms of its interpretation in the history of thought. Even a purely analytic method is necessarily linked to the historical problem, for the very character of our contemporary conception of time is but the result of a long intellectual development. This fact could suggest that time, having no reality apart from the medium of human experience and thought, is relevant only in its various historical interpretations, for no certain criteria of the truth of such interpretations can be accessible to us. True, no single and final definition of the concept of time is possible since, far from representing a unitary self-contained form, such a concept is always conditioned by man's understanding of it.[9]

In this study we focus not so much on the phenomenon of time itself but on men's and women's perceptions of it and on some of the sources of these perceptions. To begin our study, we must keep in mind that in the course of history, people have devised methods of dealing with time that in effect render it objective and measurable.[10] A clock and a calendar are devices that measure the passage of chronological time. Although two people may disagree on whether an hour is a short or long period of time, they would agree that an hour contains 60 minutes, that it can be divided into smaller units of time called seconds. Time has features, therefore, that people contend are objectively measurable.

But the concept of time also possesses subjective properties. If one asks a man if the past was good, the man may be unable to give an answer. For him the past may be filled with good and bad experiences, some continuing to please or trouble him, others forgotten. Irrespective of the actual substance of his reactions, he recognizes that the question requires him to report subjectively about certain matters, matters that he realizes cannot be objectively assessed. But his responses are made in reference to the objective features of time. If we ask this man to define an hour, we presumably will hear references to clock or calendar units.

Now, let us ask this same man to define the duration of the present. Let us even supply him with clock and calendar units—minutes, hours, days, months, and years—to help him with his answer. Again he will recognize that even with these so-called objective units of time, he cannot arrive at a purely objective definition of the present's duration. He must offer a subjective definition, a report of his *feelings* about the duration of the present.[11]

While the duration of the present is an important concern in Chapters 7 and 8, at this point we must understand the distinction between the objective and subjective nature of time perception. To do this, let us now imagine three men who have been asked the question: How long is the present? Let us also imagine that we have asked these men to choose clock or calendar units to arrive at their definitions of the present's duration. The first man reports that it should be measured in seconds or even smaller units. What he is suggesting is that only the very instant in which one experiences a sensation—for example, a pin prick—should be called the present. The instant the sensation stops, it belongs to the past. In choosing seconds or fractions of seconds to define the present's duration, the first man has attempted a definition that would be free of subjectivity and any psychological feeling states. The chronologically measurable passage of instants, which one cannot control, determines the duration of the present. Let us call this objective definition a scientific one. Let us also keep in mind that it is based on a linear conception of time. Time flows from the future (the period in time one has not yet experienced), to the present (the period in which sensation is experienced), and eventually to the past (the period when the sensation is no longer experienced.) Each day is composed of an unending line of instants in which sensations are experienced, *but only the instant of sensation itself* determines the present and its boundaries.

A second man has a different definition of the present's duration. His career, or something else that provides a sense of continuity he may not be able to articulate, cause him to feel that the present's duration should be defined in years. He may be using some standard to arrive at his decision, for example, the precise number of years he imagines he will be staying at a particular job. Like the first man, he has attempted to give an answer based on an objective calculation of the present's duration, but he has used different criteria to arrive at his answer. Let us then call this extended present an *activity*-defined present, in contrast to the first man's scientifically defined present, and let us keep in mind that while both men may have tried to objectify their responses, these so-called objective responses range from seconds or miniseconds to years.

A third man answers our question exactly as the second man answered it: the present's duration, he says, should be measured in terms of years. When asked to explain his answer, however, he can give no objective reason for his response. The best he can say is that the present's duration *feels* like years.[12] His response is wholly subjective and predicated on no calculable measure.

The point of this exercise is twofold: It demonstrates that the range of responses to a question of how long the present is will vary greatly, and it shows that people may or may not use the so-called objective features of time in answering the question. In addition, researchers who study perceptions of time may want their respondents to give fundamentally objective answers to their questions or they may encourage their respondents to give subjective reactions. Thus, responses will vary according to the researcher's intentions and methods. But all of us are capable of giving what we feel are both objective *and* subjective responses to questions concerning the meaning of time. When researchers ask their respondents to define the duration of the present, even in chronological terms, one cannot know for certain the degree to which subjective reactions will influence objective assessments. Some respondents might wish to rid their answers of anything that smacks of objectivity; others may wish to rid their answers of anything that smacks of subjectivity.

The tension between objective and subjective responses to the meaning of time is important. It raises the question: How does one perceive the passage of time? In philosophical terms, does one perceive time in a *linear* or in a *spatial* framework, the framework Henri Bergson called a durational framework? Because this distinction is of utmost importance to us throughout our study, we must take a moment to explore it.

As much as any other philosopher, Bergson believed that the passage of life was a continuous one. Because to live is to grow old, time reveals itself partly in the way one experiences one's inner life. "This inner life," Bergson wrote, "may be compared to the unrolling of a coil, for there is no living being who does not feel himself coming gradually to the end of his role."[13] Bergson was quick to assert, however, that no moment can possibly be identical to any other moment; there is always one's memory that differentiates one moment from another. I know more this minute that I did a minute ago and less than I will know a minute from now. Not even two consecutive seconds can be identical. The earlier second lacks a memory of itself which the later second will retain. As Bergson said, "A consciousness which could experience two identical moments would be a consciousness without memory."[14]

Throughout his work on the meaning of time, Bergson advocated the notion that time flows as a continuous uninterrupted line. In this way he associated himself with many physicists, particularly those who opposed the notion that time was periodic, interruptable, or that it could even be slowed down. Simultaneously he disassociated himself from philosophers like Spinoza and Descartes who had argued that the passage of time should be conceived of as atomic. According to the physicist G. J. Whitrow:

Descartes was compelled to postulate that the instants at which creaturely beings exist must be discontinuous or atomic. Temporal existence must, therefore, be like a line composed of separated dots, a repeated alternation of the state of being and the state of non-being.[15]

By postulating his theory of the continuous flow of time, Bergson was not only opposing the notion of periodicity of the flow of time, he was differentiating *temporality*, or the continuous indivisible flow of time, from *spatiality*, the conception of time as atomic and divisible. Bergson would have agreed with Van der Leeuw who warned that when one reduces a melody—time in this case—to its individual notes there is no longer a melody. By reducing music to its elementary notation one can understand it, but music requires a continuum of notes for its total effect.[16]

Bergson further argued that the inner self perceives time as a continuous flow. Moreover, this perception takes precedence over any external sense of time—a sense of time as divisible, discontinuous, or what he called spatial. His distinction between temporal and spatial time should not be confusing. In effect, Bergson believed that the word *temporal* should be used to describe the linear and continuous flow of time. When one conceives of time in a different way, one attributes spatial qualities to the passage of time.[17]

Bergson's notion of time as linear in configuration—derived exclusively from the perceptions of one's inner self—was diametric to Immanuel Kant's notion. Kant had long argued that the intuition of time and the process of representing time and change to ourselves were founded essentially in spatial terms. We might contrast Bergson's concern with the linear succession of moments with this one sentence from Kant's *Critique of Pure Reason:*

For in order to make even internal change thinkable we must represent time (the form of inner sense) figuratively as a line, and the internal change by the drawing of this line (motion) and so we are obliged to employ external perception in order to represent the successive existence of ourselves in different states.[18]

Here, then, we have an argument for examining the bearing of external factors on the way we internally sense time. But Kant has a second important aspect of time perception. Notice his phrase ". . . to represent the successive existence of ourselves in *different* states" (italics added). Implicit in this phrase is the importance of memory, the recollection of prior moments which, in the linear conception of time, can no longer be experienced.

Bergson would emphasize in his philosophy the overriding importance of memory. Present experiences, he alleged, are inevitably tied to the immediate past, to what the individual has just experienced. Indeed, the present derives meaning from the past. The physicist G. J. Whitrow expressed this same idea in the following way: "Practically we perceive only the past, the pure present being the invisible progress of the past gnawing into the future."[19]

What Kant's statement suggests however, is that memory is experienced in different ways. We can say that the minute that has just ended can never be experienced in the same way again. Although we can recall the minute, it is not the same as experiencing the minute as we experienced it originally. This is the essence of the concept of temporal succession: The past cannot be regained.

For other philosophers, however, Bergson's reasoning did not appear to capture the way in which time seems to be experienced by human beings. Friedrich Kummel argued that "succession alone cannot explain the complete essence of time."[20] A feeling we have about time may be more useful and important to us than the fact that we cannot actually recapture past moments. Memory allows us to call back prior instants, and although the recalled instants are not identical to the original instants, the fact that they are lodged in our memory means that they are not gone forever. Through memory we can recapture the past, albeit incompletely.

This point of view, arguing that memory can recapture the past, is an essential aspect of the so-called spatial or durational conception of time. It is called spatial because it suggests that instead of conceiving of time as an unbreakable chain of instants, we can use our imagination to lift a past instant out of its place on the continuum of time and drop it into another place. In the case of memory, for example, we retrieve past experiences and relocate them in the present. As some philosophers put it, we conceive of past and present instants as correlative, rather than linear and successive.[21]

To illustrate this notion of past and present moments being correlated, we may imagine ourselves at a door receiving guests. One after the other they file in, impressing their identities upon our memories. Suddenly, we are asked whether Mrs. J. has arrived and if so, whether she was wearing a coat. At this moment, still within what we call the present, we must return, through memory, to an earlier guest to regain our impression of Mrs. J. In the process of recalling her, however, our impression becomes clouded by intervening circumstances and ongoing psychological occurrences and recapturing the identical impression we had of Mrs. J. only moments ago becomes difficult. Still, we are, in effect, bringing back to the present the impression of Mrs. J., an impression caused by our meeting her in a time period we call the past.

In summary, the phenomenon of memory is important not only for an individual's perception of time, it is an important part of the scheme that philosophers and psychologists have developed in their attempt to represent the meaning of time. For what is temporality other than the perception of events that are bound to the present but must inevitably be replaced by still newer perceptions.

In the perception of any temporal phenomenon, the perceiver explains not the object but the presence and structure of the percept. The immediate impression, therefore, along with the retention of the impression, are, by definition, the immediate or present experience of perception. Husserl defined impression as a receptive act of a stimulus, or what might be called incorporation. Retention refers to the process in which just-happened events remain with the perceiver in the present but are distinguished from memory, which ultimately comes to be defined as the ability to recall information or percepts now disconnected from immediate action. In the formation of a series of retentions, an earlier retention becomes the just-perceived retention as a new retention comes to take its place in line.[22]

If memory of the past forms one basis for perceiving time, expectation of the future forms another basis. For many philosophers the concept of the future is infered from experiences in the past. Recognizing that we first experience each moment in the present and then watch it retreat into the past, we begin to realize there is something called the future, a period of time in which moments that have not yet occurred exist.

The distinction between linear and spatial conceptions also affect how one perceives the future. In the linear conception we cannot know the future because of the successive nature of moments. We must await the entry of future moments into the present to experience them. In the spatial conception of time we correlate past, present, and future moments and recognize the linearity of time and that people do become preoccupied with expectations of the future, impressions of the future, or what Husserl called "protention." In the previous illustration a question about Mrs. J. may cause us to think about her husband, to expect that he may soon arrive. In the same way we recall the impression we experienced of Mrs. J. a moment ago, we can now intuit the impression of Mr. J., not yet arrived, whose image we conjure up through memory.

Although expectation derives to some extent from memory—we have an image of Mr. J because we have met him in the past—expectation is not identical to memory. We can recall expectations we have had, but when we recall them, they are no longer expectations, but *memories* of expectations.[23] Because expectations, recollections and sensations of events occurring in the present do come into consciousness, we can demonstrate the spatial or durational conceptions of time. The spatial conception of time is based on our involvement with the past, present, and future; our belief that they are related in some way; and a feeling that there is more to time than the linear passage of succeeding moments.

In large measure, we cannot experience the phenomenon of time until it is spatialized, until we "detime" it. We merely infer what time is, even while we experience it. The line of guests arriving at a party and our impressions of these guests as well as our recollections and expectations of them reveal the fundamental tension between so-called objective, or linear, conceptions of time and the subjective, or durational or spatial, conceptions of it. This is a crucial distinction we must make in our work. In fact, it is a distinction that all people make in thinking about the meaning of time. But let us again remind ourselves that both the linear and spatial conceptions of time are relevant to us as we contemplate the meaning of time.[24] We bring temporal and spatial criteria to bear in our perceptions of time. Kummel states:

Time is traditionally described as a fragmentation of successive vanishing moments; one can, however, just as logically assert the integrity of time based on the inner correlation and coexistence of its parts. Only the two definitions taken together can fully describe the nature of time.[25]

We must not think that the two conceptions are antithetical. Instead, we must recognize that the spatial conception of time relies heavily upon the linear conception. The spatial conception does not imply that past, present, and future instants are mixed up with one another or even that they occur simultaneously. There is always a period of time that we have not yet experienced—the future—and a period of time that we have already experienced—the past. The two periods are not identical. In a paradoxical sense, therefore, the spatial conception of time rests on the linear conception.[26]

One may argue that neither conception seems relevant to the ways in which people actually experience time in their day-to-day lives. We may wish to know more, therefore, about the distinction between linear and spatial conceptions in the context of human behavior. To pursue this issue, let us turn to the results of three psychological experiments.

Years ago John Harton performed an experiment in which he asked people to find their way through a maze of alleys and corridors.[27] Actually, the maze was constructed so that the experimentor could control the successes and failures of the subjects, but they were led to believe that they alone were determining the outcome. Each person had six opportunities to traverse the maze, after which he was asked to estimate how long it had taken. Results of the experiment showed that people underestimated the time when they succeeded and overestimated the time when they failed.[28]

In a different experiment performed in 1952, Morris Eson and John Kafka asked their subjects to estimate time passage under four conditions, two of which were classified as usual, two as unusual.[29] Under what they called usual conditions, subjects were instructed to perform tasks such as silencing a continuous tone for the same period of time that they had heard it earlier. The so-called usual condition mixed the modes of stimulation. Subjects were asked to switch off a light bulb for a period of time estimated to be equal to the duration of a previously heard tone. The results of this experiment revealed that when presented with so-called unusual stimuli, subjects' estimations of time passage tended to be inaccurate. The authors concluded that an ability to recover from confusion caused by unusual stimuli is essential if perceptual processes are to be readjusted and that the ability to recover is itself a sign of emotional involvement in a task. This involvement is what finally causes variations in our perception of the duration and passage of time. The greater the involvement, the greater the variation, particularly when perceptual processes are readjusted after unusual stimuli.

With these two experiments we begin to see the ways in which our perceptions of the chronological passage of time become influenced by external stimuli, such as lights and sounds, and social psychological stimuli, such as success and failure in the performance of a particular task. We also begin to see how the objective and subjective perceptions of time interact. A third experiment makes these points even more graphically.

Using an elaborate experimental design, Jonas Langer and his associates asked their subjects to sit in an electric cart.[30] The subjects were instructed to stop the cart after five seconds, but without aid of a clock. Moving at a constant speed, the cart could be aimed at a danger goal, a precipice in this experiment, or at a safety goal, a flat surface. The major hypothesis of Langer's study was that as danger becomes imminent one overestimates the passage of time. Although the hypothesis was proved, the major conceptual value of the experiment was that differences in our objective and subjective estimations of time occur, in part, through our perception and anticipation of dangerous situations. These findings demonstrate that emotional involvement influences one's estimation of how much or how little time is taken to perform a task.

One final study that demonstrates the relationship between objective and subjective perspectives of time might be mentioned. Heinz Werner demonstrated that lights flashing in sequence appear to increase their flash interval as a function of the distance between them. The word to be stressed here is *appear*, for none of us is capable of manipulating the passage of chronological time, the time normally serving as our objective baseline in any estimation experiment. Werner's subjects could not literally manipulate the time at which the lights flashed; yet by increasing distance (or more exactly, judged duration), Werner in effect was slowing down the appearance of the interval flash by prolonging the period following the first flash. By inserting distance between the point of right now and the next now, it may *seem* as though we are altering the actual speed of time passage, but of course this is only a subjective effect.

We have now briefly reviewed the two essential conceptions of time that we will be considering in this book, the linear and the spatial. We have also examined several examples in which these two conceptions come together in the way people perceive time. We have seen how external stimuli such as light and sound, the anticipation of danger and the experience of success and failure in the performance of a task affect perceptions of time. In our review of temporal conceptions, we have examined the significance of immediate experience, the recollection of past experiences, the expectation of future experiences, and how these three types of experiences of time might relate, or in Bergson's word, correlate.

Our experiments explore the ways people perceive the temporal horizon. We are inquiring into fundamental perceptions of time and attitudes toward time, the periods of time that constitute the phases of one's life, and the historical past and future. Our goal is to understand how linear and spatial conceptions of time and personality variables, social roles, and cultural values influence our perceptions of time. In addition, we wish to know how, in objective and subjective terms, people perceive these aspects of time and what significance they attribute to them.

NOTES

1. See Kurt K. Luscher, Time: A Much Neglected Dimension in Social Theory and Research. *Sociological Analysis and Theory*, 4 (October 1974), 101–116; and Stephen Toulmin and June Goodfield, *The Discovery of Time* (New York: Harper Torchbooks, 1966).

2. See J. J. Grebe, Time: Its Breadth and Depth in Biological Rhythms. *Annals of New York Academy of Science* 112 (1962): 1206–1210.

3. See Hubert G. Alexander, *Time as Dimension and History* (Albuquerque: University of New Mexico Press, 1945).

4. "Highly developed theoretical thinking," Ernst Cassirer wrote, "tends to consider time as an all-embracing form for all change, as a universal order in which every content of reality 'is' and in which an unequivocable place is assigned to it. Time does not stand beside things as a physical being or force: it has no independent character of existence or action. But all combinations of things, all relations prevailing among them, go back ultimately to determinations of the temporal process, to division of the earlier and later, the 'now' and the 'not now.' "

   See Cassirer, *Philosophy of Symbolic Forms*, Vol. III., *The Phenomenology of Knowledge* (New Haven: Yale University Press, 1957), p. 16.

5. On this point, see M. Young and J. Ziman, Cycles in Social Behavior. *Nature* 229 (1971); 91–95.

6. Robin George Collingwood, *The Principles of Art* (Oxford: Clarendon, 1964).

7. On a related point, see C. P. Oberndorf, Time, Its Relation to Reality and Purpose. *Psychoanalytic Review*, 7 (1941): 139–155.

8. J. J. C. Smart, *Problems of Space and Time* (New York: Macmillan, 1964).

9. Friedrich Kummel, Time as Succession and the Problem of Duration. In *The Voices of Time*, J. T. Fraser (Ed.) (New York: Braziller, 1966), p. 31.

10. See Daniel N. Maltz, Primitive Time—Reckoning as a Symbolic System. *Cornell*

Journal of Social Relations, 3 (1968): 85–113; and Elizabeth Achelis, *Of Time and the Calendar* (New York: Hermitage House, 1955).

11. On this point, see Alden E. Wessman, Personality and the Subjective Experience of Time. *Journal of Personality Assessment*, 37, No. 2 (1973), 103–114; and R. Calabresi and J. Cohen, Personality and Time Attitudes. *Journal of Abnormal Psychology*, 73 (1968): 431–439.

12. On a related issue, see Bernard S. Aaronson, Behavior and the Place Names of Time. *The American Journal of Hypnosis*, 9, No. 1 (July 1966): 1–17.

13. Henri Bergson, *Matter and Memory* (New York: Humanities, 1964), p. 140.

14. *Ibid.*

15. G. J. Whitrow, *The Natural Philosophy of Time* (New York: Harper & Row, 1961), p. 156.

16. See G. Van der Leeuw, Primordial Time and Final Time. In *Man and Time*, J. Campbell (Ed.) (New York: Pantheon Books, 1957).

17. Reviewing Bergson's work, Friedrich Kummel wrote, "[He] not only rejects any fragmentation of duration into a before and after but also insists on the point that only an interrelation of all its moments may properly be considered duration. Past and future may not be dissociated from the present without destroying the unity of duration itself." See Kummel, *op. cit.*, p. 48.

18. Immanuel Kant, *Critique of Pure Reason* (New York: Dutton, 1934), Ch. ii, Section 4.

19. Whitrow, *op. cit.*, p. 87. This position of Bergson's was attacked by Russell, who claimed, according to Whitrow, that in an obviously circular approach, Bergson had in effect unwittingly removed time from his argument.

    Ernst Cassirer also reflected on this important aspect of Bergsonian philosophy when he wrote: "Action is determined and guided by the historical consciousness, through recollection of the past, but on the other hand, truly historical memory first grows from forces that reach forward into the future and help to give it form." Cassirer, *op. cit.*, p. 188.

20. Kummel, *op. cit.* p. 38.

21. See Charles M. Sherover, *Heidegger, Kant and Time* (Bloomington: Indiana University Press, 1971).

22. On this point see Philip Merlan, Time Consciousness in Husserl and Heidegger, *Philosophy and Phenomenological Research*, 8, No. 1 (September 1947). The concepts of impression and retention in phenomenological literature are not dissimilar from Bruner's notion of temporal stacking. See, for example, Jerome Bruner et al. Inhelder and Piaget's The Growth of Logical Thinking: A Psychologist's Viewpoint, *British Journal of Psychology*, 50 (1959): 363–70. The essential point in both of these theories, however, is that the consciousness of temporal events derives from the *succession* (my italics) of impressions and retentions.

23. In his theory of protentions and retentions, Husserl seemed to allege that perception of the temporal event does more than merely activate processes of the present. If each act of perception is composed of endless protentions and retentions, will not life experiences seem endless in terms of anticipations? Moreover, has he not thereby rendered being in time infinite?

24. For a personal perspective on this point, see Robert M. MacIver, *The Challenge of the Passing Years: My Encounter with Time.* (New York: Trident, 1962).

25. Kummel, *op. cit.*, p. 45.

26. Kummel summarized this point in the following passage: "Thus arises the apparent paradox that the amalgamation of time periods into a mere succession inevitably results in their mutual exclusion, whereas their distinction makes possible their harmony within an articulated unitary structural interrelation." *Ibid*, p. 44.

27. John J. Harton, The Influence of the Difficulty of Activity on the Estimation of Time. *Journal of General Psychology*, 21 (1939): 51–62. See also Harton, An Investigation of the Influence of Success and Failure on the Estimation of Time. *Journal of Genetic Psychology*, 21 (1939): 51–62.

28. These findings are consistent with reports of more recent studies that demonstrate the influences of affective states and evaluations of one's competence on time perception and specifically estimations of duration. Just as Harton's research showed that actual or anticipated success was related to an underestimation of the passage of time, so does anticipated or actual failure cause the sense of an overestimation of time passage. On this point, see John Muller, *Self, Time and Activity.* Unpublished manuscript, Department of Social Relations, Harvard University, 1967; and Muller *The Temporal Structuring of Competence.* Unpublished manuscript, Department of Social Relations, Harvard University, 1966.

29. Morris E. Eson and John S. Kafka, Diagnostic Implications of a Study in Time Perception, *Journal of General Psychology*, 46 (1952): 169–83.

30. Jonas Langer, Seymour Wapner, and Heinz Werner, The Effect of Danger Upon the Experience of Time, *American Journal of Psychology*, 74 (1961): 94–97.

31. See Heinz Werner, *Comparative Psychology of Mental Development*, Rev. ed. (Chicago: Follett, 1948).

CHAPTER THREE

ORIENTATIONS TO THE PAST, PRESENT AND FUTURE:

THE EXPERIENTIAL INVENTORY

Like heat, light and sound, time may be measured by a guage. Clocks, or calendars, or even biological and cultural cycles may be likened to thermometers, photometers, and audiometers. But helpful as clocks and calendars may be, they capture only a limited aspect of time. They do not adequately clarify subjective perceptions and spatial conceptions of time passage.

Well before the development of highly technical research skills, St. Augustine expressed the concern that one cannot be certain what aspects of time may be fruitfully studied. A logical starting point might seem to be an examination of the concepts of past, present, and future because they are the most widely used concepts. For Augustine, however, this was not a good starting point. Only the present, he said, is truly experienced. The past and future exist only in the form of memory and anticipation, and recalling and anticipating are clearly activities one experiences in the present. For Augustine, the awareness of existing in the present meant that because one does not experience time directly, one cannot measure it. Time makes an impression on the mind, and it is the impression—and only the impression—that we experience as existing in the present. As Augustine wrote:

It is in you, O my mind, that I measure time. . . . What I measure is the impress produced in you by things as they pass and abiding in you when they have passed: and it is present. I do not measure the things themselves whose passage produced the impress; it is the impress that I measure when I measure time.[1]

Many years later after reviewing a large number of psychological studies of time perception, Paul Fraisse came to almost the same conclusion about the past, present, and future:

20

In this ever-changing world our actions at any given moment do not only depend on the situation in which we find ourselves at that instant, but also on everything we have already experienced and on all our future expectations. Every one of our actions takes these into account, sometimes explicitly, always implicitly.[2]

While echoing Augustine, Fraisse also anticipated modern analysts by recognizing that to argue the existence of three distinct time zones (past, present, and future) may be misleading, particularly if human beings ultimately are able to experience these zones and events in them and feelings about them only in the present. To make his position clear, Fraisse completed a sentence with Augustine's own words, words that in effect anticipated contemporary existential thought: "The present, therefore, has several dimensions: . . . the present of things past, the present of things present, and the present of things future."[3]

Ideas about the relative significance of the past, present, and future for understanding the meaning of time have long been debated by philosophers and psychologists. To understand the differences in philosophical and psychological opinions, we must undertake a brief review of other research and writings on this topic. Because some of his ideas are now familiar to us, let us begin our review with the philosophy of Henri Bergson.

Bergson argued that the true self, as he called it, emerges when the perceptions one has in the present and influenced by recollections in such a way that one is unable to differentiate between present sensations and recollections from the past.

The true self is not the self that reaches and acts outward; it is the ego that is capable of looking back into time in pure recollection and of finding itself again in its depth. This view into the depth of time is opened up to us only when action is replaced by pure vision—when our present becomes permeated with the past, and the two are experienced as an immediate unity.[4]

Bergson believed that for knowledge of the self to exist, one must understand not only what is happening in the present but also how the past has influenced one's perceptions of the present in the present. Indeed it was Bergson who gave birth to what J. B. Priestely later called psychological time.[5] Opposing Hobbes' notion that memory stands for a secondary form of perception, Bergson argued that one perceives present sensations at the same time one reflects on prior sensations. Consideration of the future in the form of anticipation or expectation, Bergson contended, gives meaning to our lives and helps us to develop a sense of hope.[6] If, however, we want to understand the nature of living in time, we must examine the process of memory. In the words of Marcel Proust:

Those who have created for themselves an enveloping inner life pay little heed to the importance of current events. What alters profoundly the course of their thinking is much more something which seems to be of no importance in itself and yet which reverses the order of time for them, making them live over again an earlier period of their life.[7]

Bergson's thesis, that the true nature of time is experienced essentially through the act of memory and can be understood only through an examination of memory, might now be contrasted with Alfred North Whitehead's doctrine of mutual imminence:

The understanding which we want is an understanding of the insistent present. The only use of knowledge of the past is to equip us for the present. *The present contains all that there is* [my italics]. It is holy ground for it is the past and it is the future.[8]

Whitehead reaffirmed this position elsewhere: "Again a vivid enjoyment of immediate sense data notoriously inhibits apprehension of the future. The Present moment is then all in all."[9]

In Bergson, and perhaps in Augustine as well, we have found advocates for what might be called a *past dominance*. Both men believed that to understand the experience of time, one must understand the infiuence of memory on immediate perceptions.[10] Whitehead, however, stands as an advocate of what we might now call a *present dominance*. Two other advocates of this present dominant position are Edmund Husserl and Martin Heidegger, whose distinctive phenomenological theories suggested that one's consciousness of time exists in one's sense of the present.[11] Reminiscent of what Augustine had propounded, both Husserl and Heidegger stated that while we experience the "impress" of present events and sensations, neither recollections of the past nor anticipations of the future can determine the quality of these impressions.[12] Husserl and Heidegger alike clashed with Bergson on this thesis of historical time. Bergson believed historical time was inseparable from the acts of recalling and perceiving immediate sensations. The phenomenologists argued that knowledge, the essential intuitive understanding of events existing in history, could not be authentically experienced. The significance of this point is readily made. Every person recognizes the transitory nature of his or her own time—that is, his or her finitude. The only two things we know with certainty are that we are and that eventually we no longer will be. But—and this is essential to understanding Husserl's reasoning—we can know these things neither through appearance nor through *direct* experience, but only *indirectly* through the experience of others, literally through their reports. Our knowledge of a finite self derives, therefore, from interaction, involvement, and inference. From the point of view of phenomenological theory, I can never experience my birth or death; hence, I can never perceive either event as temporal, as I can never know (phenomenologically) their beginnings or ends.

The third logical position, what we might call a *future dominance*, is found in the writings of Ludwig Binswanger and Rollo May. Although these two theorists subscribe to the unity of being and time, established in part by Bergson, they emphasize not only the significance of anticipation or expectation of the future, but the *developing* or *becoming* nature of our experiences of time. May, for example, wrote " . . . that the more crucial fact about existence is that it *emerges*—that is, it is always in the process of *becoming*, [emphasis added] always developing in time, and is never to be defined at static points."[13] Similarly, Binswanger declared that " . . . the primary phenomenon of the original and authentic temporality is the *future* [my italics], and this future in turn is the primary meaning of existentiality, of the designing of one's self 'for one's own sake.' "[14]

A similar position in which the future is considered the most important time zone, was derived by George Kelly from a clinical psychological context:

Behavior is given its consistency by attempts to anticipate events. The particular behavior in which an individual engages presumably reflects the anticipation he has for the future, and his anticipations are expressed in his constructs.[15]

Kelly's position seems to differ markedly from Bergson's position about the true self and its relation to memory. Essentially, Kelly states that the experience of time is future-oriented. Time is constantly moving toward the future, and while we may well be experiencing momentary sensations or recalling prior sensations, our major life activity is preparing for the future. This preparation occurs through the acts of anticipating or expecting. Because we cannot return to a prior time through any effort of our own, we must use the present to prepare for the future. According to

Kelly, our experience of time may be likened to walking. Regardless of the direction we choose, we are always facing the future. Even if we walk backward, we still face the future. If we turn around in hope of retracing our steps, little is accomplished beyond establishing a new future course.[16] As Ernst Schachtel said, "Walking requires faith in the future."[17]

Even this brief philosophical and psychological review shows how views on the past, present, and future and their significance for understanding how we perceive time tend to differ. From our experiences in everyday living, we might argue that to understand our perceptions of time, we must consider all three time zones equally. We all know the experience of recalling the past, of watching events happen in the present, and of anticipating other events in the future. But where individuals—and societies and cultures—differ is in the significance they attribute to each of the three time zones as they attempt to understand their own perceptions of time.[18] This is the major point we should draw from the foregoing review.

Indeed, when Florence Kluckhohn and Fred L. Strodtbeck observed, "The possible cultural interpretations of the temporal focus of human life break easily into the three-point range of Past, Present, and Future,"[19] their intention was not to suggest that the people of a particular culture act in accordance with their feelings about only one time zone. Rather, they wished to suggest that we can categorize cultures (and even individuals perhaps) according to the value cultures place on past-, present-, and future-oriented activities—activities such as remembering, or experiencing immediate sensations, or anticipating future sensations. As Kluckhohn and Strodtbeck said:

Obviously, every society must deal with all three time problems; all have their conception of the Past, the Present and the Future. Where they differ is in the preferential ordering of the alternatives (rank-order emphases), and a very great deal can be told about the particular society or part of a society being studied and much can be predicted about the direction of change within it if one knows what the rank-order emphasis is.[20]

To suggest, therefore, that a culture or a person is oriented toward the future means only that a great value is placed on doing "future-oriented" tasks, such as saving money for the future or even predicting the future. It does not mean that the culture or person is totally unaffected by recollections of the past or by experiences in the present. It also does not mean that one's recollections of the past and one's experiences in the present play no part in one's expectations of the future. Surprisingly, some social scientists theorize about people as being, say, future oriented, as if to suggest that these people consider only the future to be important to them and their lives. Actually, the goal of such theorizing is to establish categories or typologies of time perceptions. Our own goal in this chapter is to develop a typology based on the significance men and women accord to the past, present, and future—men in one way, women in another. In building this typology, however, we must remember not only the significance of memory, the perception of immediate sensations in the present, and the expectations of the future; we must also be aware that men and women deal with all three time zones differently and simultaneously in the course of living.

THE EXPERIENTIAL INVENTORY

The method we have used to build our typology is based on the placement of important experiences in time. We asked the hospital corpsmen and corpswomen to examine past and present experiences and their expectations of future experiences and to tell us which experiences are to them most important. The instrument we used to examine these experiences is called the Experiential Inventory. The instructions for the first half of the Inventory read as follows:

Please list the ten most important experiences of your life. These may be experiences that you have had, you are having, and experiences you expect to have. You only need to write a few words for each experience. You may list your experiences in any order you wish.

The purpose of the second half of the Experiential Inventory was to locate each experience in a particular time zone. The instructions read as follows:

Now that you have listed 10 experiences, please study the time zones below:
   Time Zones
   1. Distant Past
   2. Near Past
   3. Present
   4. Near Future
   5. Distant Future
Take each experience and decide if it has occurred, is occurring, or will occur. Then choose the number of the time zone that best represents the time of the experience

and write this number on the dotted line in front of the experience. Do this for all ten experiences.

The instructions did not indicate that a person could not assign several numbers to any and all experiences. In fact, we expected that some persons would list all five numbers by experiences they could not locate in any one zone. In such instances we computed a mean.

If numbers are assigned to all 10 experiences, an average number or mean score may be computed by adding the numbers and dividing by 10. This mean score suggests a general location of a persons experiences in time: The higher the score, the more a person's experiences were oriented toward the future, the lower the score, the more they were oriented toward the past.

In using the Experiential Inventory we assume that people are free to range over the entire temporal horizon, from the distant past to the distant future, to find and then place their 10 important life experiences. Perhaps unless an experience has acutally occurred, it cannot be considered important. If so, fewer future expriences and more past and present experiences would be identified on the horizon. However, if the act of anticipating experiences is in itself a major component of human behavior, as Kelly asserted, many future experiences should emerge on the Experiential Inventory. At least the importance to the hospital corpsmen and corpswomen of the future should emerge in the form of some near- and distant-future experiences. Our main assumption, therefore, is that the Experiential Inventory provides a way to examine the significance that people attribute to the past, present, and future. If a person appears to be especially preoccupied with experiences that occurred in one time zone, we may then say that this person seems more oriented to this zone than to another zone.[21]

To repeat, the Experiential Inventory yields two main pieces of information: a mean score representing the general location in time of a person's 10 important experiences and an actual count of the number of experiences located in each of the five time zones. However, these data do not merely signify the person's orientation to the temporal horizon or identify his or her place on the temporal horizon; they also indicate the ways in which memory, the perception of present experiences, and anticipation of future experiences influence people as they identify important life experiences.

Before examining the results of the corpsmen's and corpswomen's performance on the Experiential Inventory, we should make one final point relevant to all the results we will be considering in our study.

An orientation to a particular time zone, revealed by the point on the horizon where one places important life experiences, does not imply any other sort of time perception. People maintain different perceptions of time and these perceptions need not relate to one another. A person deeply involved in planning for the future or one who only articulates expectations may have an equal desire to recover the past or indulge in recollection. We do not know whether the person who seems preoccupied with the future conceives of the future as an important or unimportant time zone. We only know where in time people are locating their ten most important experiences.

Results. Men and women show almost no difference in where they locate their 10 experiences. The experiential means are 2.34 and 2.31 respectively.[22] These values fall at the lower end of the range, a point located between the near-past time zone and the present time zone. Thus, the students cited more recalled experiences than anticipated experiences.

The mean scores on the Experiential Inventory do not imply that the students considered future experiences irrelevant, or not worth speculating upon, or unimportant. If we examine the distributions for the Inventory (Table 1)*, we find that most of the students located at least some experiences in all five time zones, and only a few students put all experiences in one zone.

If this occurred strictly by chance, we would have found two experiences in each zone and most of the people would have had a mean score of 3.0—a perfect present orientation. However, we do not see a chance distribution. First, the two past time zones are overrepresented, while the remaining three time zones are underrepresented. Second, over 68% of the sample listed one or no future experience. Third, about 55% of the sample failed to place even one experience in the distant future.

If we look at the experiences themselves, we note two major types. First, the *discrete* experiences, so called because they exist primarily in one time zone. An example of a discrete experience would be a person's year in kindergarten. Although a significant part of his or her later development may be traced to what happened in this one year, the person perceives kindergarten as finite, limited to one past period or time.

The second type of experience is called *diffuse;* it encompasses experiences of the past, present, and future.[23] To demonstrate the difficulty a

*All tables appear at the end of the book.

person has in trying to locate a so-called diffuse experience in one time zone, consider the following example: A woman in the first part of the Experiential Inventory lists "education" as one of her 10 most important experiences. Following instructions in the second part of the test, she is now obliged to locate education at some point in time—that is, she must choose from her entire educational experience the period that holds greatest meaning. If she thought grade school most important, she would probably designate some past time zone. If she felt that education should be characterized as a continuing moment-by-moment learning experience she would assign it to the present, although she could also assign it to the future.

It did not happen often, but several students evidently did have trouble placing certain experiences in only one time zone. Instead, they chose all five zones. An example was "loving my parents." On his first try, one young man chose a 3 to reflect his present love. Believing that he had excluded his past love for his parents, he went on to add 1 and 2—that is, the distant past and near past. He then crossed out these numbers leaving a score of 1–5 as his final decision.

Two additional findings emerge from the data. First, perhaps because they were dissatisfied with their present school situation, men identified fewer significant experiences in the present than women. The difference amounted to about 18%. While there is little evidence from other research to suggest that men generally are more dissatisfied than women with their present circumstances, studies show that women accept the demands of school more easily than men. Possibly this accounts for the descrepancy.[24]

The second finding is that the number of present experiences (see Table 2) was negatively associated with the number of distant-past experiences (male $r = -.29$, $p < .01$; female $r = -.23$, $p. < .05$) and near-past experiences (male $r = -.24$, $p < .01$; female $r = -.29$, $p < .01$), but relatively unrelated to the number of near future experiences (male $r = -.06$, n.s.; female $r = .03$, n.s.) and distant future experiences (male $r = -.13$, n.s.; female $r = -.16$, n.s. These data suggest that a focus on the past tends to exclude interest in the present and that a focus on the near future can coexist with a present orientation or exist without it. More generally, experiences were viewed as being associated with either the *real* world of ongoing experiences or with experiences as they are *imagined*. That is, people concentrate either on the real experiences that involve them in the present instant or on their recollections of past experiences or anticipations of future experiences.

But so-called imagined time can be further refined. As we have stressed, there is a great difference between recollection and expectation. Recollection "returns" to the present something that once was very real. Expectation, of course, doesn't because we have not yet experienced the future.[25] Whatever the psychological process of choosing important experiences and placing them in time zones, we distinguish ongoing present experiences from both recalled and expected experiences. Whitehead appears to have been correct when he said, "The present, then, is all in all."[26]

Summary. The Experiential Inventory instructed the corpsmen and corpswomen to scan the completed, continuing, and anticipated experiences of their lives and to select the 10 most important ones. Completed or recollected experiences tended to equal in number anticipated and continuing experiences. There is a clear negative relationship between focusing on the past and focusing on the present. Present and anticipated experiences, however, were unrelated. Furthermore, while the directions of the statistical correlations shown in Table 2 could be influenced by the Experiential Inventory's scoring procedures, the values of the correlations are sufficiently low, suggesting that the relationship among recollection of past experiences, perception of continuing experiences, and the expectation of future experiences is not traceable to a flaw in the scoring procedures we used. Although all the students were constrained in their thinking by the instructions of the Inventory, the minor variance found in the correlations should not be attributed to the nature of the Experiential Inventory alone. We might have believed that because they are young, the corpsmen and corpswomen would choose to deemphasize the importance of their past experiences and concentrate on experiences in the future, but just the opposite occurred. By asking these people to list the 10 most important experiences of their lives, we essentially asked them to search among completed or ongoing experiences.

CONCLUSION

By reviewing the writings of several authors, we pointed out that there is disagreement over which time zone is most significant for our understanding of the nature of time. We placed these authors into categories represented by the time zone each author felt was most significant. Because he insisted that memory is the most important concept for understanding time, we categorized Bergson as "past-dominated;" because he believed that all mental activities, recollections, sensations, and anticipations exist in the present we categorized Whitehead as "present-dominated"; because he stressed the constant development of the personality and the need people have to plan for the future, we categorized Kelly as "future-dominated."

Following these three points of view and guided by Kluckhohn's and Strodtbeck's reminder that one need not be totally committed to a particular time zone to have some special but limited attachment to this time zone, we presented the results of the Experiential Inventory. In building a typology of temporal attachments or orientations to time zones based on the location in time of significant life experiences, we cautioned that any typologies that categorize people on the basis of their perceptions of time must not convey the idea that people maintain simple and unchanging perceptions of time. The danger in performing research of the type described here is that labels like past-, present-, or future-oriented may start out as ways to conceptualize how we perceive time, but often conclude as rigid categorizations of human experience. Although they help us to organize large groups of people into useful classifications, single-zone designations like "lower-class-present orientations" or "middle-class-future orientations" also distort the way people perceive time.[27] People possess complex orientations to, and perceptions of, the temporal horizon, and we must be careful not to ignore the subtleties of this complexity. Writing about the intricacies of perceiving the past, present, and future, Lawrence Frank made clear the difficulty most of us encounter merely in reflecting about time:

Seemingly then, the future determines the present, the present controls the past, but the past creates the future and so imposes its values on the present. Is this but juggling of terms and obscurantism or does it reflect the actual processes of human conduct that our traditional teaching of philosophy, ethics and much of psychology have heretofore obscured under the pretentious conception of man as rational, volitional and autonomous?[28]

Frank's comments stimulate us to consider a wealth of temporal issues, including the phenomenon of duration, the forward and backward extension in time we perceive in a particular time zone.

Although the results of the Experiential Inventory help us develop a typology to see how people relate to time, the results cannot be fully understood until we examine the way the corpsmen and corpswomen perceived the five time zones and their durations and the way they located their experiences in these times zones. Suppose, for example, that one person calls a period that starts five years from today the distant future, a second person calls it the near future, and a third person calls it the present. If all of these people are placing significant experiences into the same period but if they are perceiving the period in three different ways, three different orientations to time are developing from experiences located at precisely the same chronological point in time. This problem—the perceived duration of the distant and near past, the perceived duration of the present, and the perceived duration of the distant and near future—is the subject of the next chapter.

NOTES

Quoted in Ernst Cassirer, *Philosophy of Symbolic Forms*, Vol. III, *The Phenemenology of Knowledge* (New Haven: Yale University Press, 1957), p. 168.

2.   Fraisse, *The Psychology of Time*. New York: Harper and Row, 1963, p. 151.

3.   *Ibid.*

4.   Cited in Cassirer, *op. cit.*, p. 185.

5.   J. B. Priestley, *Man and Time* (Garden City, N.Y.: Doubleday, 1964).

6.   See Bergson, *Time and Free Will* (London: Allen and Unwin, 1910).

7.   Cited in Albert William Levi, *Philosophy and the Modern World* (Bloomington: Indiana University Press, 1959), p. 72.

8.   Harold B. Dunkel, *Whitehead on Education* (Columbus: Ohio State University Press, 1965), p. 99.

9.   Alfred North Whitehead, *Symbolism, Its Meaning and Effect* (New York: Putnam's Sons, 1959), Capricorn Books, p. 42.

10.   See J. A. Adams, *Human Memory* (New York: McGraw Hill, 1967); and F. C. Bartlett, *Remembering: A Study in Experimental and Social Psychology* (Cambridge: Cambridge University Press, 1932).

11.   See G. A. Berger, Phenomenological Approach to the Problem of Time. In *Readings in Existential Phenomenology*, N. Lawrence and D. O'Connor (Eds.) (Englewood Cliffs, N. J.: Prentice-Hall, 1967), pp. 148–204.

12.   On this and related points, see Anna-Teresa Tymieniecka, *Phenomenology and Science in Contemporary Human Thought* (New York: Farrar, Straus & Giroux, 1962).

13.   Rollo May, Contributions of Existential Psychotherapy. In *Existence*, Rollo May, Ernst Angel, and Henri Ellenberger (Eds.) (New York: Basic Books, 1958), p. 66.

14.   Ludwig Binswanger, The Case of Ellen West. In *Existence*, R. May, E. Angel and H. F. Ellenberger (Eds.) (New York: Basic Books, 1958), p. 302. The force of this theme of temporal emergence lies in the significance of an inevitable future state for existence. The difficult writings of Heidegger describe the nature of this force (italics mine): "If the term understanding is taken in a way which is primordially existential, it means to be

projecting towards a *potentiality for being* for the sake of which any/being/exists. . . .
When one understands oneself projectively in an *existential possibility*, the future
underlies this understanding, and it does so as a *coming-towards-oneself* out of that
current possibility as which one's being exists. . . . Factically, (being) is constantly
ahead of itself, but inconstantly anticipatory with regard to its existential possibility."
Martin Heidegger, *Being and Time* (New York: Harper & Row, 1962), pp. 385–386.

15. In Lee Sechrest, The Psychology of Personal Constructs. In *Concepts of Personality*,
Joseph W. Wepman and Ralph W. Heine (Eds.) (Chicago: Aldine, 1963), p. 211.

16. It is interesting to note Ernest G. Schachtel's words: "Walking requires a faith in the
future." Schachtel, *Metamorphosis* (New York: Basic Books, 1959), p. 48.

17. On a related point, see L. S. Dickstein and S. J. Blatt, Health Concern, Futurity and
Anticipation. *Journal of Consulting Psychology*, 30 (1966): 11–17.

18. See Marian W. Smith, Different Cultural Concepts of Past, Present and Future.
Psychiatry, 15 (1952): 395–400; and Arden R. King, Time, Society and Culture. *Human
Mosaic*, 3 (1968): 1–11.

19. Florence R. Kluckhohn and Fred L. Strodtbeck, (Eds.) *Variations in Value Orientations* (Evanston, Ill.: Row, Peterson, 1961), p. 13.

20. *Ibid.*, p. 14.

21. We should not forget, however, that the Inventory in a sense stills these important
experiences by creating artificial qualities of temporal duration and succession.

22. The standard deviations are 6.46 and 6.21 respectively.

23. Cf. C. Gordon, Self Conceptions: Configurations of Content. In *The Self in Social
Interaction*, C. Gordon and K. J. Gergen (Eds.) (New York: Wiley, 1968).

24. Jerome Kagan, Acquisition and Significance of Sex Typing and Sex Role Identity.
Review of Child Development Research (New York: Russell Sage Foundation, 1964).

25. The exhilaration of an expectation, as well as its ominous qualities are captured in this
passage by Meerloo: "Man has a special delight in anticipation. It is a typical masochistic
reaction in which one pays the price for desire in advance. The glamor of anticipation
contains, as I have said, the fantasies of the happy past, but at the same time, it contains
also many unhappy occurrences transformed by time into happy ones, and these involve
various past guilts that have never been solved or cleared up. These mental doubts too
can now be paid off in advance or on the installment plan through the blanket coverage of
a generally happy future making up for everything." Joost A. M. Meerloo, *The Two
Faces of Man: Two Studies on the Sense of Time and on Ambivalence* (New York:
International Universities Press, 1954).

26. Note that his division of temporal experiences into so-called real and imagined ones is
somewhat different from the division made by Medard Boss. Boss argued that one is
existentially confronted with "here and now's" and "then's," which can ultimately relate
either to the past or to the future: "Original time is no external framework consisting of
an endless sequence of 'now' on which man can eventually hand up and put into proper
order his experiences and the events of his life. . . . Man's original temporality always
refers to his disclosing and taking care of something. Such original temporality is dated
at all times by his meaningful interactions with, his relating to, that which he encounters. Every now is primarily a 'now as the door bangs,' a 'now as the book is missing,' or a
'now when this or that has to be done.' The same holds true for every 'then.' Originally, a
'then' is a 'then when I met my friend sometime in the past' or a 'then when I shall go to
the university again.' " Medard Boss, *Psychoanalysis and Daseinanalysis*. Tr. by L. B.
Lefebre (New York: Basic Books, 1965), p. 45.

27. See Edward Banfield, *The Unheavenly City* (Boston: Little, Brown, 1968); also R.A. Ruiz, R. S. Revich, and H. H. Krauss, Tests of Temporal Perspective: Do They Measure the Same Construct? *Psychological Reports*, 21 (1967):849–852.

28. Lawrence K. Frank, Time Perspectives. *Journal of Social Philosophy*, 4 (1939): 293–312.

CHAPTER FOUR

THE PERCEIVED DURATION OF TIME ZONES:

THE DURATION INVENTORY

A major concern of our study is the way people perceive time and the way in which they represent it to others. Do they measure the passage of time objectively, using chronological standards, or do they measure it subjectively, according to the way they *feel* about time? Or do they measure it both ways? How do they view or define such things as the distant or near past, the present, or the near and distant future? What do these time zones really mean to them?

The essence of the last question is captured in another question: How long does a particular time zone last? When does it start and when does it end? To answer these questions, people may use chronological units, such as seconds or microseconds, or they may use cultural units, such as the duration of a holiday. Or they may use no standard units at all. They may say that the present seems to last for years. They may measure it subjectively by their feelings.

Despite the enormous number of references to the past, present, and future in reports of psychological research, few people have undertaken systematic studies of perceptions of the basic time zones.[1] John Cohen, who did undertake such a study, demonstrated that when time is represented by a straight line, younger people see the period of a day as being longer in duration than older persons see it.[2] Yet even Cohen's study is not really addressed to the issue of how we perceive the duration of time zones.

The philosopher Eugene Minkowski divided the temporal horizon into seven zones, each with its own unique conceptual and experiential value: distant, mediate, immediate past; distant, mediate and immediate future; the present.[3] Yet even in this elaborate division of the temporal horizon we still find no understanding of how people perceive nearness or distance, nor where they place the boundaries between time zones. Nor do we know anything about the perceived duration of the present.

The task of inquiring into the perceived duration of time zones is important for several reasons. First, we refer to time zones in the course of everyday language, implying that we do have some sense of their duration. Second, how people perceive the duration of time zones provides us with a method for examining the degree to which linear and spatial representations depict various aspects of time. Once again, we recall the distinctions illustrated by our hypothetical example of the three men estimating the duration of the present. Although two of the men used chronological units and still arrived at different estimates of how far the present extended, they nevertheless *did* use them. The third man also used years to make his estimate, but when questioned further, he indicated that his estimate was based on a spatial conception of time, not on a linear conception. He was not concerned with the passing of successive moments; he was concerned with the way he *experienced* the period of time called the present rather than with finding the best chronological unit to describe its duration.

Our research turns to the question: How do people perceive the duration of time zones? We wish to know how long the present lasts and what are the boundaries that separate the distant past from the near past, the distant future from the near future, and the present from both the near past and near future.

COMPUTATIONS OF DURATION

In a study by Morris Eson and Norman Greenfield, respondents were asked to list what came to mind when they thought about the near past.[4] Eson and Greenfield were then able to demonstrate that most of these thoughts and impressions were characterized by anticipation rather than by recollection. For example, respondents mentioned work they were planning to finish. Throughout the discussion of their findings, however, the authors never questioned the way *they themselves* defined the near past—in this case, the two preceding weeks.

In a family of studies involving attitudes toward the future, investigators collect lists of people's expectations, hopes, intentions—even descriptions of the way people see themselves several years in the future. While these techniques do lead toward revealing important aspects of one's personality, they do not help us to understand how, in chronological terms, people perceive the duration of the future. Why is it that when one asks 18-year-old boys with similar social class backgrounds, occupations, and intelligence, to list their plans for the future, some of them see the future as being a few weeks hence, others a decade?[5]

In our use of the Experiential Inventory students were instructed to list the most significant experiences of their life, then to note where each experience occurred in time: in the distant past, near past, present, near future, or in the distant future. Although these zones helped the respondent organize his or her experiences and provided a way to help us quantify all test results, the Experiential Inventory failed to define the boundaries of the five time zones. When two people said all their experiences occurred in the distant future, they could be classified together as future-oriented people. But if we find that one person perceives the distant future in terms of months while the other perceives it in terms of years, the single category, future-oriented, is vague and imprecise and not very helpful in understanding the nature of people's involvement with the future.

Two people may designate the identical experience, the birth of a child two years ago, as belonging to different time zones. To one person, two years ago might be the distant past; to the other, it might be the present. On the Experiential Inventory, the first person would mark a 1 to represent the distant past; the second person would mark a 3 to represent the present. Thus, two different numerical scores would be given to the same experience.

By using chronological units to define the duration of time zones, we are examining the passage of time in both spatial and linear terms. We are examining people's perceptions of a sequence of moments, which implies a linear conception of time, and then how these moments relate to one another, which implies a spatial conception. The distinctions between linear and spatial conceptions of time are important. As Van der Leeuw points out, summarizing Bergson's notions of temporal duration (le temps-durée):

In reality there are no segments. Nor is time homogeneous; one second passes continuously into the next and is likewise inseparable from those that preceded it. Real time is a river [after Heraclitus], a melody. One can count seconds, because they are not time but space. In temps-durée there is no counting, any more than one can dissect a melody into notes.

To ask people to define the duration of a time zone in clock or calendar units, we are asking a deeper question: On the basis of your experiences and feelings, how long does a time zone seem to last?

In specifying how long the present is, some people would say it is the precise instant one experiences a particular sensation—feeling a pin prick or hearing a door bell ring, for example. It begins when the sensation starts, ends when the sensation stops; the length or duration of the sensation equals the length or duration of the instant. The sensation provides boundaries for the instant; the instant is the present, and the present is defined strictly in terms of physical sensation.

A person might also define the present's duration as the smallest *measureable* unit of time, presumably some fraction of a second. Here the definition is based on an objectively measurable passage of time. In both cases linear measurement has been employed and hence the determination has been based on objective calculations.[7]

We may use the period of time during which something happens to conceptualize the present's duration a third way: action duration. For example, if we define the present as the period of time consumed by a thought or a speech, we might conceptualize the present's duration in terms of minutes or hours.[8] Although larger units of time could be employed in the definition, the flow of time remains continuous, at least during the period the happening is taking place.[9] The fundamental conception of time implied in this definition is beginning to change from the strictly linear and is becoming spatial. One indication that the action-duration conception is becoming a spatial conception is that persons may experience earlier and later feelings[10] or feelings of stopping and starting anew.[11] All these feelings are seen as belonging to the present. If, however, one conceives of the present as precisely the instant of a sensation, one must then define the past and future as the times before and after the sensation. Because they embody the boundaries of the present, the terms *before* and *after* or *earlier* and *later* by definition cannot actually be *in* the present, at least when the present is defined in sensate and scientific terms.

In the action-duration definition of the present one may conceive of a series of earliers and laters that do not serve as present boundary markers; instead, they are part of the present. Now the present is constituted of things past and future,[12] things that are different from those outside the boundaries of the present. This is what is meant by a spatial conception of time, one that is not confined to the ineluctable forward flow of moments.

A fourth scheme of the present's duration is clearly derived from a spatial conception of time passage. In *compound-action* time the present may last as long as months or years. Compound-action time denotes an extended present whose duration is influenced by sociological and cultural definitions of periods, cycles, or epochs.[13] One speaks, therefore, of a *present* occupation or *present* world conditions. Indeed, one may define the present as the time extending from one's birth to one's death. Pesumably, the society and the culture influence one's personal definition.

Although different amounts of time are employed in the last three definitions, the present is seen in periods of time extending well beyond the moment of sensation.[14]

Summary. The duration of the present ranges theoretically from the instant of sensation (or of the scientific microsecond), through longer periods in which personal action or effort takes place, to still longer periods determined societally or culturally. Each of these schema reflect different experiences of, or involvements with, time; each treats the passage of time in a somewhat different manner. At various points in their lives people may find that any one or all four schemes will illustrate the way they perceive the duration and boundaries of the present. Bergson wrote:

There is no doubt but that for us time is at first identical with the continuity of our inner life. What is this continuity? That of a flow or passage, but a self-sufficient flow or passage, the flow not implying a thing that flows, and the passing not presupposing states through which we pass; the thing and the state are only artificially taken snap shots of the transition; and in this transition, all that is naturally experienced is duration itself.[15]

With each shift of feeling a person may change his or her conception of the duration of the present or, for that matter, of any period of time. Moreover, when the extension of the present is changed, extensions of the past and future are also changed.

THE DURATION INVENTORY

To examine how the corpsmen and corpswomen conceptualized, chronologically, the five time zones presented in the Experiential Inventory—the distant and near past, the present, and the distant and near future—we used the Duration Inventory.

The instructions read as follows:

In the following sentences fill in the blank spaces with one of the words listed below. Indicate your choice by placing the number of the word in the blank space.

Select only one word for each blank space.

1. Second
2. Minutes
3. Hours
4. Days
5. Weeks
6. Months
7. Years

1. The *present*, as I think of it, extends from _____ ago to _____ from now.
2. As I think of it, the *distant past* includes things and events which occurred
   _____ ago while the *near past* includes things and events which occurred
   _____ ago.
3. As I think of it, the *distant future* includes things and events which will occur
   _____ from now, while the *near future* includes things and events which will
   occur _____ from now.

Although the Duration Inventory encourages respondents to divide the horizon into three portions (past, present, and future), some persons make their computations for one zone only after taking into consideration how they defined the duration of other zones. For example, they might extend the present from the moment where the near past leaves off to the moment where the near future begins. Some persons may also respond to each of the statements of the Duration Inventory one by one and thereby not consider the duration of one zone as they compute the duration of another zone. By filling out the Duration Inventory in this manner, they might end up with what we call a *discontinuous temporal horizon* in which one time zone may not begin immediately after the previous time zone has ended. For example, the present ends days from now, but a near future begins months from now. Between these two zones is a timeless space; hence, we call the conception discontinuous

Because our respondents could select any unit of time to complete the Duration Inventory, various outcomes are possible, for example, a present that began seconds ago but will not end until years from now. We call this configuration in which the boundaries of the present are measured in different units *an asymetrical present* and contrast it with a *symetrical present* in which both the present's boundaries are measured in the same units.

Results.   Table 3 charts responses to the Duration Inventory. It discloses that men and women tend to hold similar perceptions of nearness and distance but vary somewhat in how they perceive the duration of the present. Men show more of a tendency than women to shorten the present. The widest discrepancy between the sexes is found at the future boundary of the present (13%). Men show that they tend to see the present as a succession of backwardly flowing moments, from the present to the near past; 42% of them would say that the very next instant that lies in store is properly called the future, as opposed to reporting that the future does not start until next year.

Generally, the data concerning the present's duration arrange themselves into a bimodal distribution. At the one end are those who reveal an instantaneous present and, because of this, a prolonged past. At the other end are those who reveal an extended present, thus, a postponed future. Little variation is found in attitudes toward nearness and distance. Most students agreed that distance was best computed in years and nearness in months. Slight variations, common to men and women, suggest that for some people the near future seems closer to the present than the near past. In a sense, these perceptions of personal time coincide with the backward flow of chronological time and with human activity generally.[16]

Despite the large percentage of people who measured the present's duration in brief units of minutes and seconds, few measured the near future that way. If the present ends seconds from now, as 40% of the men report that it does, the near future should remain just out of reach, perhaps minutes from now. But only 4% of the men reported perceiving it that way. Indeed, the unit most commonly chosen to measure the length of the near future was months. Presumably, because students took each sentence of the Duration Inventory one by one, the Duration Inventory had something to do with this. In addition, the structure of the statements contained in the Duration Inventory may have reinforced previously existing perceptions of time. Individuals may simply have perceived the time zones as unrelated. Although the Duration Inventory does not provide sufficient evidence to discuss this matter fully, another instrument, the Circles Test, does (Chapter 7). Here we might anticipate the issue of the discontinuity between time zones by recalling the words of Samuel Beckett:

Yesterday is not a milestone that has been passed, but a day stone on the beaten track of the years, and irremediably part of us, within us, heavy and dangerous. We are not merely more weary because of yesterday, *we are other*, no longer what we were before the calamity of yesterday.[17]

Asymmetric Time: The Metaphor of a Present in Advance of Itself. That the near past seems farther away in time than the near future was reported by 13% of the students. The month was the unit most often chosen to measure the near past's duration; minutes, days, and weeks were used to measure the duration of the near future. We now begin to see a pattern. The boundary separating the present from the past was perceived to be thicker than the boundary separating the present from the future. Chronologically, there seemed to be more time between past and present zones than between present and future zones. Interpreted another way, this suggests a present in advance of itself, a term reminiscent of Georges Gurvitch's conception of "time in advance of itself" or "time pushing forward." "[This] is a time," Gurvitch wrote, "where the discontinuity contingency and the qualitative triumph together over their opposites. *The future becomes present*" [emphasis added].[18]

The phrase "The future become present" is essential. The present in advance of itself connotes a present that, almost as one experiences it, becomes separated from the near past, while at the same time it becomes connected to the near future. The future becomes present, as Gurvitch suggested, because one perceives it as being located *in* the present. One zone, in this case the future, resides within another zone, the present. This perception of time is based on a spatial conception of time flow, one that deemphasizes the idea of the succession of moments. The perception was discussed in philosophical terms by Ludwig Wittgenstein in *The Brown Book:*

There is the idea that the feeling, say, of pastness is an amorphous something in a place, the mind, and that this something is the cause or effect of what we call the expression of feeling. The expression of feeling, then, is an indirect way of trans-mitting the feeling.[19]

The results of the Duration Inventory make it seem as though the students in our study were more concerned with their anticipations of the future than they were with their recollections of the past. Their perceptions of the duration of time zones suggest that they bring the future closer to the present but keep the past separated from the present.[20]

Arithmetically, the present in advance of itself was found by subtracting scores representing the present's future boundary from scores representing its past boundary. The results of these computations are shown in Table 4. About 60% of the population revealed values equal to zero. These people then demonstrated a symmetrical present. Regardless of how long the present was perceived to be, the same chronological unit was selected to represent and measure both the past and future boundaries. In addition, 34% of the women and 26% of the men revealed a present behind itself— —that is, a present where the present-past boundary is thicker than the present-future boundary. Both perceptions are complex. The present in advance of itself imparts a perpetually just-starting quality to the present, while the present behind itself imparts a just-ending quality to the present. In Wittgenstein's words, these perceptions are "gestures of pastness, futureness," or in this particular case, "presentness." Hence, they may well be an "elaborate and exact expression of a feeling of pastness (or presentness) . . .".[21]

Possibly because of their youth, the students in our sample who perceived a present in advance of itself saw the future as not only lasting longer in chronological terms than the past, but they judged the future as being more important to them subjectively than the past. When a time zone diminished in importance, the size or perceived duration of the time zone was also diminished. This would be logical for young people who have so much of their lives before them. But the issue of time zone importance is complex and merits a discussion of its own in a later chapter.

Duration necessarily implies not only how long time zones last, but how fast they succeed each other. Although we did not ask people how fast a time zone passed them by, we may still assume that duration and speed are correlated. Because a minute is shorter than a year, a year seems to pass more slowly than a minute. Similarly, the present in advance of itself suggests that future experiences enter the present more slowly than present experiences enter the past. Conversely, in the present behind itself, the future enters the present more quickly than present experiences leave the present for the past. The most popular conception, the symmetrical present, suggests that the speeds are the same: The future enters the present and the present enters the past at the same rate.[22]

This notion suggests several things. One is the effect of familiarity and pleasure on the perception of the passage of time.[23] We know that happy periods move more swiftly than somber ones. The ride to an unfamiliar place seems to take longer than the ride back. In a similar way the future remains new, and no matter how much we may plan for it, it still seems to take more time arriving—more time, that is, than it seems the present takes to become the past.

EXPERIENTIAL ORIENTATIONS AND THE DURATION OF THE PRESENT

One way to explain the present in advance of itself is to compare the location of experiences in time (on the Experiential Inventory) with the duration of the present (on the Duration Inventory). We might predict that students who showed a future orientation on the Experiential Inventory would say that the future comes more quickly. Past-oriented students who relied more heavily on recall in choosing their 10 important experiences would say it comes more slowly. We might also predict that future-oriented people would perceive the boundary between the present and future to be thinner, while past-oriented people would perceive it to be thicker.

An alternative hypothesis might also be developed, based on the relationship between how people perceive the future and how they perceive the past. Some people may feel a tension between their desire to shape their own future and a belief that they must honor the past and the traditions of their culture.[24] To be consciously expectant of the future is also to be unconsciously enthralled with the past. We might predict, therefore, that future-oriented people should tend to see the boundary between the present and future as thicker than the present-past boundary, thus delaying the future's arrival.

To analyze the relationships between the Experiential and Duration Inventories, responses to both inventories had to be categorized. Normally, this procedure could be done simply by dividing respondents into two or three groups on the basis of whether they scored high, medium, or low on each of the Inventories. In the Experiential and Duration Inventories, however, the process of categorizing respondents is more difficult. Not only did the students show variations in their choice of experiences and in locating their experiences in time, they disagreed on the chronological durations of the five time zones provided in the Experiential Inventory. Moreover, because most students reported a large number of past experiences, we were prevented from simply dividing the sample into three equal groupings based on mean scores (the lows, the mediums, and the highs). Doing this might have accentuated the skewness of the sample toward the past. Finally, because those who emerged as future-oriented turned out to be special cases, they had to be examined separately from people with especially low scores.

Taking all of these factors into consideration, we still were able to divide the scores on the Experiential Inventory into some sort of groupings. The first group included people who had scored from 1.0 to 1.9—the *past-oriented* people. As reported in Chapter 3, persons might score in the 1.0 to 1.9 range in different ways. They might have had two experiences designated as distant future and eight as distant past, producing a mean of 1.8. Being past-oriented, therefore, does not imply that an individual has not listed *any* future experiences.

The second group included those whose mean scores ranged from 2.0 to 2.9. Although this range is still slightly past-oriented, it does represent a middle range of scores. This group was called *present-oriented*.

The third group included those who scored from 3.0 to 5.0. This group, called *future-oriented*, represents the people with the highest scores on the Experiential Inventory. But note again that being future-oriented does not rule out the possibility that the respondent assigned experiences to the near or even distant past, few though they were.

The distribution of the three groups is summarized in Table 5. As the mean values suggest, there were no significant sex differences within groups, although men showed a slightly higher (percentage) representation among the past-oriented group, and women among the present-oriented group. No sex differences occurred within the future-oriented group. In Chapter 5, where we explore relationships between orientations to time and perceptions of the future, this group is more significant.

Responses to the duration of the present on the Duration Inventory were also divided into three groups. At one extreme were those who conceptualized the present in terms of seconds and minutes, a conceptualization we call the *instantaneous present*. At the other extreme were those who conceptualized the present's duration in terms of months and years, a conceptualization we called the *extended present*. A third group included the people who selected the units hours, days, and weeks. This group was called *middle*.

After computing the statistical relationships among the groups, we found that future-oriented women extended the future boundary of the present to the point where they in effect were postponing the future's arrival ($\chi^2 = 8.53, p > .10$). This finding is shown in Table 6. The table also shows that future-oriented women extended the past boundary of the present as well ($\chi^2 = 5.61, p > .10$). Once again, men's responses to the Duration Inventory remained unrelated to the responses they gave on the Experiential Inventory.

The duration of the present shown by future-oriented women resembled the duration of the present shown by past-oriented women more than the present duration of present-oriented women. The present-oriented women quickened the passage of the present by shortening the present at both its past and future boundaries.[25] Thus, the term *present orientation* already begins to seem misleading.

Given the nature of the Experiential and Duration Inventories and the pattern of results presented thus far, we can reasonably argue that future-oriented women put many experiences in the present merely because to them the present appeared extended. This is only partially true. There was only a slight tendency for future-oriented women to locate more of their experiences in the present, less of a tendency for past- or present-oriented women to do it. ($\chi^2 = 3.32$; n.s.). However, 60% of the past-oriented men failed to list any present experiences at all ($\chi^2 = 25.38; p < .001$).

An examination of the number of near-future and distant-future experiences listed by women shows that future-oriented women were significantly higher in distant-future experiences ($\chi^2 = 28.98, p < .001$). Although future-oriented women extended the present toward the future, they did not locate a significant number of experiences in the present, nor did they put their experiences in the near future to any statistically significant degree. Instead, they placed a high number of experiences in what they called the distant future. Presumably, then, future-oriented women are involved with a portion of time—the distant future—they cannot actually experience. Yet, by shaking off the past and prolonging the present, they seem to postpone the arrival of the distant future, the very zone in which they have located their most important experiences.

TIME EXPERIENCE AND THE DURATION OF THE NEAR AND
DISTANT FUTURE

Do future-oriented people, specifically, extend the near- and distant-future zones? This would be difficult to determine, considering that over 80% of our sample perceived the distant future in terms of years. Previously, we observed that for men, the duration of the present was not related to where in time they located their important experiences. Now we wish to know how these men perceived the duration of the near and distant future, the zones containing so many of their important experiences. The results in Table 7 show that future-oriented men significantly lengthened the near future ($\chi^2 = 8.92$, $p < .10$).[26]

Because the future of future-oriented women is filled with so many important experiences, we might argue that these women will significantly increase the length of the future on the Duration Inventory. However, we might also argue that past-oriented people will lengthen the distant future, because to them the distant future appears slow in arriving. Although the evidence in Table 7 is scant, there is some reason to believe that for men and women, both past and future orientations tended to be associated with a lengthened distant future. The outstanding finding, however, is that future-oriented men viewed the lengthened distant future in terms of years ($\chi^2 = 25.55$, $p < .001$). Involved with their expectations for the future, they perceived the distant future as an especially protracted period. The same relationship emerged for women, but not at a statistically significant level ($\chi^2 = 3.82$, n.s.).

CONCLUSION

To be rated present-oriented on the Experiential Inventory, one must group most of his or her experiences in the present, balance each past experience with a future experience, or do both. But as we saw, present-oriented women did not overload the present with experiences; instead, they diffused them throughout the five time zones. This finding suggests that a present orientation does not imply a preoccupation with the present; rather, it indicates concern for past and future experiences as well. Nonetheless, a present-oriented person on the Experiential Inventory will have a shortened temporal horizon.

The complexity of these data should discourage investigators who classify people according to simple categories of past, present, or future. We have evidence that men and women vary in their perceptions of the temporal horizon—specifically, they perceive time zones to have different durations. Let us recall that Eson defined near past for his respondents as the period of the past two weeks.[27] Although to compare Eson's sample of men with our own sample of men is unfair, it may be worthwhile to point out that about 22% of the men in our study put this same two-week period in the *present*, not in the near past. Do the corpsmen perceive time differently? Perhaps. More important, can we determine a person's orientation to past, present, or future on the basis of a list of important experiences? Although some present-oriented people perceived a short near and distant future, they did not necessarily wish for the future to arrive quickly. For future-oriented men, a quickly arriving future is almost unheard of: only 3% of the sample measured the distant future in units as brief as months.

Results of the Duration Inventory have made one point abundantly clear: People perceive the duration of time zones differently. This becomes greatest for the present. For this zone, the corpsmen and corpswomen used all the chronological units they were given in the Duration Inventory to make their estimations. The perceptual variations also confirm what we suggested in our illustration of the three men who disagreed on the duration of the present. Our respondents were not jotting down numbers thoughtlessly. This is confirmed by the statistically significant relationships between results of the Experiential and Duration Inventories.[28]

A person's orientation to time is associated with his or her perception of the length of a particular time zone, and the length attributed to one time zone may relate to the length attributed to a second time zone. In examining the facts we are able to see patterns emerging among a series of subjective perceptions of time. These patterns are not always the same for men and women. Although not many differences by sex occur in the data we have examined so far, toward the end of our study these differences become a major topic of discussion.

One of the crucial questions in our research is what determines people's perceptions of time. What, in other words, are the sources of perceptions? In later chapters we look at some of these sources, such as cultural values, social roles, and certain aspects of personality. Throughout our study we also explore the effect one perception of time has on another perception of time. We have already begun this by examining the relationships between the Experiential and Duration Inventories.

If we look closely at these relationships, and particularly those that involve perceptions of the past and future, we sense that not only are people employing spatial schema in their responses to the Experiential and Duration Inventories, but they also ignore a great many chronological features of time altogether. To respond to questions about the past and future arouses in people fantasies of recovering the past and knowing the future.

NOTES

1. Cf. Melvin Wallace and Albert I. Rabin, Temporal Experience, *Psychological Bulletin*, 57 (1960): 213–236; and Stephen Klineberg, Changes in Outlook on the Future Between Childhood and Adolescence. *Journal of Personality and Social Psychology*, 7 No. 2 (1967): 185–193.

2. John Cohen, Subjective Time. In *The Voices of Time*, J. T. Fraser (Ed.) (New York: Braziller, 1966), pp. 257–75; and J. Cohen, C. E. M. Hansel, and John Sylvester, An Experimental Study of Comparative Judgments of Time, *British Journal of Psychology*, 45 (1954): 108–114.

3. Minkowski, Findings in a Case of Schizophrenia Depression. In R. May, E. Angel and H. F. Ellenberger (Eds.), *Existence* (New York: Basic Books, 1958), pp. 127–138.

4. Morris E. Eson and Norman Greenfield, Life Space: Its Content and Temporal Dimension. *Journal of Genetic Psychology*, 100 (1962): 113–28.

5. Thomas J. Cottle and John Muller, *An Empirical Investigation of the Intention*. Unpublished manuscript, Department of Social Relations, Harvard University, 1967.

6. G. van de Leeuw, Primordial Time and Final Time. In *Man and Time*, J. Campbell (Ed.) (New York: Pantheon, 1957), p. 326.

7. Bergson was well aware of this not so subtle transition of temporal thought into spatial terms. "But if our science thus attains only to space," he wrote, "it is easy to see why the dimension of space that has come to replace time is still called time. It is because our consciousness is there. It infuses living duration into a time dried up as space. Our mind, interpreting mathematical time, retraces the path it has travelled in obtaining it." Henri Bergson, *Duration and Simultaneity, with Reference to Einstein's Theory*. Tr. by Leon Jacobson (Indianapolis: Bobbs-Merrill, 1965), p. 60.

8. Jean Piaget, Time Perception in Children. in Fraser, *op. cit.*

9. On this point, see Martin Heidegger, *Being and Time* (New York: Harper & Row, 1962).

10. J.M.E. McTaggart, Time. In *The Philosophy of Time* R. M. Gale (Ed.) (Garden City, N.Y.: Doubleday Anchor, 1967).

11. See van der Leeuw, *op. cit.*

12. Saint Augustine, *Confessions*. Tr. by J. F. Pilkington (New York: Dial, 1964).

13. On this point, see Mircea Eliade, Time and Eternity in Indian Thought. In *Man and Time*, J. Campbell (Ed.) (New York: Pantheon Books, 1957); and M. Eliade, *Cosmos and History: The Myth of the Eternal Return* (New York: Harper & Row, 1959); Wilbert E. Moore, *Man, Time and Society* (New York: Wiley, 1963); Pitirim A. Sorokin and Robert K. Merton, Social Time: A Methodological and Functional Analysis. *American Journal of Sociology*, 42 (March 1937): 615–629.

14. On a related point, see Philip G. Zimbardo, Gary Marshall, and Christina Maslash, Liberating Behavior from Time Bound Control: Expanding the Present Through Hypnosis. *Journal of Applied Social Psychology*. 1, No. 4 (1971), 305–323.

15. Bergson, *op. cit.*, p. 44.

16. Friedrich Kummel, Time as Succession and the Problem of Duration. In Fraser, *op. cit.*

17. Samuel Beckett, *Proust* (New York: Grove Evergreen, 1957), p. 3.

18. Georges Gurvitch, *The Spectrum of Social Time* (Dordrecht, Holland: D. Reidel, 1964), p. 33.

19. Ludwig Wittgenstein, *The Blue and Brown Books* (New York: Harper & Row, 1958), p. 185.

20. Robert Kastenbaum, Time and Death in Adolescence. In *The Meaning of Death* H. Feifel (Ed.) (New York: McGraw-Hill, 1959), Ch. 7.

21. Wittgenstein, *op. cit.*, p. 184.

22. One cannot say that those students demonstrating a symmetrical present perceive it as being particularly long or short, for the symmetrical present is associated with both instantaneous perceptions and extended ones.

23. Cf. William James, *Principles of Psychology* (London: Macmillan, 1891).

24. Fred L. Strodtbeck, Family Interaction, Values and Achievement. In *Talent and Society*, D. McClelland *et al.* (Eds.) (Princeton, N.J.: Van Nostrand, 1958).

25. By extending the present's duration, future-oriented women made it more difficult for themselves to score high on the Experiential Inventory. With the present prolonged, it is harder to locate many experiences in the near and distant future and thereby become labeled future oriented.

26. The same pattern held true for future-oriented women, although the statistical results did not reach significant levels.

27. Morris E. Eson, *An Analysis of Time Perspectives at Five Age Levels.* Unpublished Ph.D. dissertation, Committee on Human Development, University of Chicago, 1951.

28. As the Duration Inventory "picked up" after the Experiential Inventory to demonstrate subjective variations in time zone definitions, a new instrument ideally now should "pick up" after the Duration Inventory so that individual unit subjectivities might be explored. Perhaps one might ask respondents to express the speed of unit passage in miles per hour. Total lifetimes or pasts, presents, and futures might also be "measured" with such a methodology, but note again how these expressions capitalize on spatial perspectives of time.

CHAPTER FIVE

FANTASIES OF TEMPORAL RECOVERY AND KNOWLEDGE OF THE FUTURE:

THE MONEY GAME

The distinctions between linear and spatial conceptions of time passage are mainly between the objective and subjective perceptions. In the linear or objective conception, we *know* about our experience of time. In the spatial or subjective conception we *feel* about it. Once experienced, a moment moves into the past, never to be experienced the same way again. Still, by recalling the past we return it to the present. Thus, we *feel* we can reexperience the past.

Feelings can take precedence over knowledge. This happens when we fantasize about time, when our wishes govern our perceptions. The wish to recover the past and the wish to know the future are two such fantasies.

Recovering the past and knowing the future are fantasies most of us have experienced. They belong to the realm of imagination. Although we know these fantasies cannot become facts because we know objectively that time passes, we entertain them, making ourselves happy or sad by doing so.

Both fantasies involve totally unrealistic ways of dealing with time, but fantasizing about the past is different from fantasizing about the future.[1] The difference is important. We can actually recall certain moments from the past or even wish that we could have a bit of the past back, but we do not imagine the future that was. We cannot "recall" it. We may expect certain events will occur, and we can fantasize future happiness or wish to know the future now; but nothing in the future can be known with the same certainty that we know events of the past. Because we have not experienced the future, our fantasies about it are free of our experiences of it. Thus fantasies about the past and future are as different as recalling or redoing and planning or hoping.

Herbert Fingarette captured the vividness of time fantasy with his notion of "sense of presence":

The sense of "presence," of nearness in subjective time is generated, then, when any object or situation is cathected by the currently mobilized drives and when it plays a significant role in the dominant drive-fantasy complex. Such a "dynamic theme," or "unifying theme," once mobilized in waking life or in the dream, is like a magnet....The current perceptions incorporated are then perceived as "real"; the memories, though perhaps locatable in long past (calendar) time, are, in subjective time, "as if it happened yesterday"; the hopes are vividly present: "I can see it already!"

Generalizing from Fingarette's statement, one senses a freedom in fantasy, a freedom that William Henry said "derives from the less conscious and less structured aspects of the individual's personality. To these areas of personality the rules of logic and propriety do not apply."[3] We are free to fantasize about the past and future as we may fantasize about the historical past, the period of time preceding our birth, and about the historical future, the period of time following our death.

Fantasies about time may or may not build upon other people's reports and conjectures. Thus we may distinguish between *authentic fantasies*, those developed in part from memory or expectation of personal experiences, and *inauthentic fantasies*, those evolving from secondary sources and experiences. These we have never known directly nor will we ever know them.[4]

THE MONEY GAME

On the Experiential and Duration Inventories our respondents were reporting their perceptions and attitudes about very real aspects of time. In the Experiential Inventory they placed (or located) important experiences into zones on the temporal horizon; in the Duration Inventory, the corpsmen and corpswomen were schematizing these zones and estimating their length. To inquire into the fantasies of recovering the past and knowing the future we developed a test instrument called the Money Game. From it we hoped to learn the answers to two questions: How willing are people to fantasize recovering time from their personal and historical pasts, and how willing are they to "preknow" time in their personal and historical futures?

The Money Game is a measure of pure fantasy. The first word of the Money Game's instructions confirms this:

Pretend that you have a lot of money, more money than you can possibly use. Pretend also that someone has the power to sell you time, any time that you want, and you have this time given to you right now, knowing what you now know. And once you have this time, you can do whatever you wanted with it. Now look at the following amounts of money: (a) $0, (b) $10, (c) $100, (d) $1000, (e) $10,000. For each of the questions below, circle the letter that represents the amount of money that you would pay. Make certain that you answer all 12 questions.

How much would you pay to right now bring back:
1 hour of your own past?   a b c d e
1 day of your own past?   a b c d e
1 year of your own past?   a b c d e

How much would you pay to right now know all about and do whatever you wanted with:
1 hour of your own future?   a b c d e
1 day of your own future?   a b c d e
1 year of your own future?   a b c d e

How much would you pay to bring back right now any:
1 hour before you were born?   a b c d e
1 day before you were born?   a b c d e
1 year before you were born?   a b c d e

How much would you pay right now to know all about and do whatever you wanted with any:
1 hour that will occur after your death?   a b c d e
1 day that will occur after your death?   a b c d e
1 year that will occur after your death?   a b c d e

The idea for the Money Game came from a study conducted by Leonard Doob on how long people will delay gratification. In the study Doob asked his respondents whether they would rather have a certain amount of money now or more money at some specified future time.[5] In this research money and what people would do with it seemed a useful test device, one with clear meaning to the respondents and one that could provide the researcher with a way of quantifying their responses. However, a task like the one in the Money Game may present some people with certain problems. To have an endless amount of money to play with may perplex some people, which could alter the outcome of their test performance.[6] In addition, the fantasy of recovering past time may be more threatening to some people than we might expect. Some may not wish to recall their pasts; hence the Money Game could arouse unpleasant feelings which could interfere with test performance.[7] Although fantasizing about the past may be threatening, Joost Meerloo has indicated that fantasies about the past are an important part of one's conscious and unconscious life:

> Here it may be noted that in all inner disturbances, the time factor is a cardinal point. There is always primarily a search for past time, for the obscure and forgotten crisis or the might-have-been; it is an attempt at recapturing it and working it out differently, usually more happily, or for simply dwelling on it. We are most of us amateur time detectives, groping blindly for the clues to our own mishaps and errantries.[7]

> And it may be, in some cases, that he sees the past as having been altogether happy, and he wishes to repeat it in the future, minus the errors that brought that happy past to an end—and of course, the whole thing will be larger and more significant. Or, on the other hand, the past may have been miserable to him, and he wishes to turn it inside out and make a happy future of it—constituting a triumph, a vindication, and also a kind of satisfactory revenge on life.[8]

Similarly, fantasies of knowing the future may also be threatening. Generally, these fantasies represent three distinct wishes. First is the wish to have the future arrive quickly. Thus, we use the phrase, "I can hardly wait." Second is the wish to determine the outcome of one's present efforts. Knowing the content of even an hour of one's future would provide powerful evidence of one's worth and achievement—or one's lack of them. Third, and perhaps most threatening, is the wish to postpone death. To know even an hour beyond death means that one lives past one's time and achieves immortality.

Joost Meerloo examined the fantasy of knowing the future in the context of the disturbed personality:

> A note on the problem of precognition is needed in a study of subjective time because in every neurosis the craving for timelessness and foreknowledge of future occurrences plays a role. We may explain this as an unconscious wish for immortality and eternity. It is, at the same time, a reaction against dissatisfying reality, because perceptions of time in an adult, in contrast with the magic infantile omnipotence which ignores time and schedule, means a necessary confrontation with actual reality and its limitations. Often, when a person announces his foreknowledge, his vision of the future, one can explain this vision as wishful thinking, as illusion caused by the patient's unconscious need to put himself beyond time, to evade the pains and stresses of waiting and uncertain anticipation.[9]

The Money Game is based partly on an individual's reaction to money, specifically, on the manipulation of money. People could respond to the financial values, rather than with feelings toward time zones. The prediction that the more time respondents are offered, the more money they will be willing to spend for it is logical. In addition, while the Money Game may intimidate some people, others may not take it seriously. Because of its premise of limitless wealth, they might spend all their money on all of the units of time. Upon reading the instructions, one young scientist predicted exactly this. As we are about to see, however, he was wrong.

Results. In examining the way spending styles differed between men and women we might have predicted that any task involving the expenditure of large sums of money would appeal more to men than women, if only because in our culture major financial operations are usually controlled by men. Moreover, if spending money in some way enhances a man's sense of his own power, we might assume that in the Money Game men would spend more than women. These predictions, however, are only partially supported by the data. Only at the higher financial values ($1000 and $10,000) did men demonstrate a greater predisposition than women to invest in time.

Men and women alike offered some of their unlimited funds for time from both their personal past and future. The opportunity of recovering as much as a year discouraged some people who preferred to get on with real living. Still, about 50% of the students were interested in purchasing at least some of their personal pasts and futures. At this point the Money Game had aroused some interest in our respondents. However, when it came to purchasing their historical pasts, more than 70% of our respondents refused.

Recovery of the Personal Past. The frequency distribution of payments in the Money Game for the personal past is shown in Table 8.[10] A desire to recover a year of their personal past was demonstrated by 67% of the men and 46% of the women. Refuting the doubting social scientist, unlimited wealth did not provoke unlimited spending, but the prediction that people would spend more money for more time was confirmed. A positive relationship emerged between unit of time and sum of money: The more time we offered our respondents, the more money they offered us. Because of this relationship, offering different units of time might seem to be unnecessary. We might have offered our respondents only years, omitting hours and days. In fact, certain variations in responses to the Money Game can be traced directly to the units themselves. Significantly, we found that although more men than women wished to recover time from their personal pasts, men wanted a year whereas women preferred an hour or a day. This difference caused the mean payment scores between the sexes to differ.

As Meerloo suggested, a clue to why men prefer temporal recovery more than women is found in the fact that men want to take their recovered time and rework it. In posttest interviewing men reported that once they had witnessed the outcome of their earlier plans, they wanted to experiment with different life plans and circumstances. Having first fantasized the retrieval of time from their past, they then fantasized changing it, thus changing the present and future course of their lives.

Women's use of recovered time was very different. In posttest interviewing they reported a desire to relive past time. Women essentially bought past time to recapture their prior experiences and feelings. The fantasy of recovering a certain amount of time from the past meant that they could relive old experiences without having them affect their present or future lives.

Preknowing the Personal Future.   Unlike the fantasy of recovering the past, the fantasy of knowing the future appealed almost equally to men and women: 41% of the men and 47% of the women refused to pay anything to know in advance a year of their personal futures (Table 9). In fact, combining the results shown in Tables 8 and 9 indicates that about 40% of the students remained unwilling to play the Money Game and to fantasize about their personal past and future.

Recovery of the Historical Past.   Although people cannot actually recover their past or know what their future holds, they have at least experienced a past (the personal past), and presumably they will experience a future (the personal future). The Money Game asks them to fantasize about both. When we focus on *historical* time, however, fantasies become inauthentic because the historical past cannot be regained by recollecting one's own personal experiences. Nor are the contents of the historical future inferred in the way that we infer the contents of the personal future. As we noted earlier, the important distinction between authentic and inauthentic fantasies is that an authentic fantasy is about our own life events, and an inauthentic fantasy is about events that have not occurred or will not occur in our own life. We may imagine how we might have lived had we been born 300 years ago or how our lives might look if we were born 300 years from now, but in imagining these scenes we know that we are dealing with events that we can never directly experience.[11]

Our interest in the historical past may be stimulated by several things. We may be intellectually involved with a particular period of history and think it would be a congenial period to live in for a while, or our interest may stem from a desire to trace the origin and course of ourselves or our family. For Karl Mannheim, the historical past provided knowledge of growth and evolution. Contemporary existence develops from a past and within it a past shared in various degrees by all societies.

This idea of "the past which lies behind" can thus be interpreted in two ways: as a temporal past or as an antecedent evolutionary phase which can account for any particular detail of the actual. Looked at from the former point of view, everything has meaning because it has arisen out of a temporal process of development; from the latter point of view everything that exists historically has meaning because it exhibits the same fundamental drive, the same basic trend of mental and spiritual growth.[12]

A more personal statement or involvement with the historical past is contained in the following passage by Edward Tiryakian.

Another aspect of the openness of the self lies in its historical grounding; being-in-the-world means to participate in social time and thus in history. I am today what I am in part because of my historical past and part because of what I anticipate to be my historical future. I am also historical in a collective or social sense, that is, I am open to and take as mine the history of my people and this leads me to realize that I am not contained in my finite and solid appearance but that my being goes out spatially and temporally.[13]

That a concern with the historical grounding of one's life dominates all our experiences with time is doubtful. For example, young people given formal religious training may develop a curiosity about history, but they would probably not develop a passion to recover time from the historical past. The distribution of payments for the historical past shown in Table 10 corroborates this. Less than 30% of the corpsmen and corpswomen thought that time from the historical past was worth recovering. This result should seem remarkable to the social scientist who predicted that if respondents were given unlimited amounts of money, they would spend it indiscriminately. More important, the fact that buyers were so few indicates that the time zone in question, in this case the historical past, determines whether people want to play the Money Game. Neither the unit of time nor the fact that they have unlimited funds sways people.

The argument that a person would have to be concerned with the complexities of existential or evolutionary questions to be interested in the historical past does not explain this lack of interest in fantasizing about recovering time from the historical past. One might just wish to observe one's parents or grandparents before one was born. One interpretation of the lack of interest is that people feel themselves to be detached from the events in this particular time zone. Apparently, they are unable to propel themselves backward in time or to use their fantasies to develop an understanding of their present lives or present world conditions. Even the promise of unlimited wealth is insufficient.

Preknowing the Historical Future.   Despite all the distinctions we may make between and among authentic and inauthentic fantasies and linear and spatial conceptions of time, the future still cannot be known or experienced until it arrives. The closest we can come to knowing the future is to imagine it. As Maurice Merleau-Ponty wrote:

Reproduction presupposes re-cognition and cannot be understood as such unless I have in the first place a sort of direct contact with the past in its own domain. Nor can one, *a fortiori*, construct the future out of contents of consciousness: no actual content can be taken, even equivocally, as evidence concerning the future, since the future has not even been in existence and cannot, like the past, set its mark upon us.[14]

Still, imagining what the personal future holds can be distinguished from imagining what the historical future holds. If one knew the contents of the historical future, one would be able to evaluate such a thing as existence itself or the evolution of cultures. It would also be equivalent to gaining immortality. To know the historical future is to eliminate death.[15] While this may seem obvious, the way our respondents reacted to this possibility in the Money Game is less obvious.

From a philosophical point of view, we should recall that for Husserl, death was not a moment in time and hence not a temporal event. Similarly, Heidegger wrote, "The future is the ability to endure oneself in one's finiteness."[16] Putting it as Hegel did, "Man pays the price of his individuality by being capable to contemplate death.[17]

The following citation from Cournot, selected by Miguel de Unamuno, introduces still a different aspect of the historical future: destiny. The passages accompanying the citation from Unamuno's own work depict the religious quality of destiny and hence the religious quality of the historical future:

Religious manifestations are the necessary consequences of man's predisposition to believe in the existence of an invisible, supernatural and miraculous world, a predisposition which it has been possible to consider sometimes as a reminiscence of an anterior state, sometimes as an intimation of a future destiny.

And it is this problem of human destiny, of eternal life, or of the human finality of the Universe or of God, that we have now reached. All the highways of religion lead up to this, for it is the very essence of all religion.

Once again I must repeat that the longing for the immortality of the soul, for the permanence, in some form or another, of our personal and individual consciousness, is as much of the essence of religion as is the longing that there may be a God.

And nevertheless men have not ceased endeavoring to imagine to themselves what this eternal life may be, nor will they cease their endeavors so long as they are men and not merely thinking machines....Man will never willingly abandon his attempt to form a concrete representation of the other life.[18]

In the Money Game the historical future interested more men (49%) than women (29%) (Table 11). Furthermore, the distribution of payments again indicates that people play the Money Game on the basis of their interest in a time zone, in this case the historical future, rather than on the basis of the unit of time offered.

Summary. We may now summarize performance on the Money Game by examining the mean scores of payments for all the time zones shown in Table 12. Not only did men pay more than women for each unit of time, but there was a positive relationship between how much they bought and how much they paid for it. The more time they wanted, the more money they paid. Based on mean scores, the four time zones may be ranked in order of how much the students invested in them: Personal past first, followed by personal future, historical future, historical past. The students fantasized less about historical time and more about personal time.

One minor deviation in Table 12 deserves mention because it tends to support a recurring finding in the posttest interviews. In discussing their desire to recover personal time, many women admitted that while the idea appealed to them, a year seemed too large an amount of time to recover. The offer to recover as much as a year caused them to feel that their own lives would somehow be adversely affected if they recovered too much time from their past. Thus, women preferred in their fantasies to recover smaller amounts of time, such as an hour or a day.

That men generally paid higher amounts on the Money Game suggests that they are more ready to resolve real problems through fantasy. Or if fantasies of recovering the past and knowing the future seem to obstruct realistic life goals and if they thus become annoyances, men may prefer to keep these fantasies from interfering with ongoing realities. Again, merely the unlimited money and the chance to play with it may have influenced the way men performed.

The results of the Money Game reveal certain differences in the way men and women performed, but for the most part nothing conclusive can be drawn from them. However, in recovering the past and knowing the future through fantasy, men and women alike evaluate their fantasies concretely—that is, in terms of real present life circumstances. They are influenced, therefore, by external factors, such as the nature of a job or position on a career line, and these factors either encourage or discourage their willingness to fantasize.

The Game's instructions may have confused some of the students. The instructions encouraged respondents to fantasize about time, but they also suggested that the time recovered or known could be used in any way. The implication is that time retrieved or known through fantasy will change respondents' lives somehow. One may be reluctant to play the Money Game, not because one cannot fantasize, but because one cannot reconcile the conflicting demands of reality and fantasy. For some students, the way they construe their lives may make it difficult for them even to imagine fantasizing about past and future time, or at least admit to themselves that these fantasies are tempting.

THE RELATIONSHIP OF PAST TIME RECOVERY AND EXPERIENTIAL ORIENTATIONS TO TIME

By examining the way people perceived time in the various tests and the way their perceptions interrelate, we can also see the way so-called internal sources of time perception relate to each other. Examination of relationships between the Experiential Inventory and the Money Game are indicative of internal sources of time perception.

We recall from Chapter 3 that respondents were divided into three groups according to their scores on the Experiential Inventory. Past-oriented persons showed mean scores ranging from 1.0 to 1.9; present-oriented persons, from 2.0 to 2.9; and future-oriented persons, from 3.0 to 5.0.

Respondents may also be divided into three groups based on their performance on the Money Game. In the first group are those who refused to pay anything to recover the past or know the future. Because they are refusing the opportunity to fantasize, we call these people *realists*. Persons who offered money, no matter how little, for at least some time in a particular time zone are the *fantasizers*. We can subdivide this group into two smaller groups: persons who offered *low* amounts ($10 or $100) and those who offered *high* amounts ($1000 or $10,000) for a particular unit of time.

To investigate the relationships between the Experiential Inventory and the Money Game, we consult a set of cross tabulations between the six groups: the past-, present-, and future-oriented groups from the Experiential Inventory and the realists, low-paying, and high-paying fantasizers from the Money Game. The cross tabulations reveal that men's willingness to fantasize about recovering the past or knowing the future were essentially unrelated to their experiential orientation to time—that is, whether they were past-, present-, or future-oriented. Experiential orientations however, were related to women's performances on the Money Game.

Fewer (20%) present-oriented women played the game of fantasy than future-oriented women. When the unit of time in the Money Game was changed from an hour to a year game playing among all women decreased, but the greatest decrease was among future-oriented women. Hardly anything was offered for a year of a time zone by 29% of future-oriented women who previously had offered moderate amounts for an hour of the same time zone. A small amount of the past or future was one thing, apparently, but recovering or knowing an entire year was more problematic. Although a large majority of future-oriented men offered at least some money for a year of time from their personal pasts, 27% of the future-oriented women refused to offer any money at all for a year of their own personal pasts.

We may attribute these discrepant findings to at least two factors. First is a sex difference. Future-oriented men do not performing like future-oriented women. Second, time units themselves affect the relationships between experiential orientations to time and game performance.

THE RELATIONSHIP OF KNOWING THE FUTURE AND EXPERIENTIAL ORIENTATIONS: THE TIME OF REFRACTION

In considering relationships between experiential orientations and fantasies of knowing the future, we might reason that future-oriented people are more preoccupied with death than past- and present-oriented people because they have invested more in the future. A future orientation implies the expectation that one will live long enough to realize his or her plans. If this reasoning is correct, the threat of death would seem intrinsic to the experiential future orientation. This argument could also be reversed: People who are most preoccupied with death orient themselves toward the safety of past experiences.[19] Either way, future-oriented people are concerned with carrying experiences to their conclusion and thus should be eager to know the historical future, not only because they may wish to postpone their death but also because they may wish to know how their plans and expectations will work out.[20] This is not to say that present- or past-oriented people fear death any less than future-oriented people; an experiential future orientation may encourage a sense of incompleteness or lack of fulfillment to a degree that a past orientation may not. Older past-oriented persons may even say, "I can't complain; I've had my time."

The fantasy of knowing the future is complex. In addition to understanding the distinctions between the personal and historical futures and between authentic and inauthentic fantasies, we must know the sorts of cultural and religious traditions that shape conceptions of the future—for example, the Calvinist tradition that asks men and women to toil in the present for future rewards. We must also be aware that the consideration of the historical or personal future arouses in people the need to reconcile their present aspirations with the inevitability of death.

In thinking about how the corpsmen and corpswomen might view the future, given some knowledge of it, to derive benefits in the present, let us be certain to distinguish the fantasy of knowing the future from normal expectation or anticipation.

Perhaps the best way to understand the fantasy of preknowledge is through the notion of *refracted time*. In the concept of refracted time the present is perceived and evaluated in terms of the future. Thus, if I feel that my work at this time is not going as well as it did a year ago, my perception of the present is in part shaped by my memory of events that occurred a year ago. If I say, however, that I am already feeling guilt, the same guilt I should soon feel when my work is done and I look back and ask, "Why didn't I do it better?" I am, in effect, experiencing the *present* in a *future* context. I have used expectation to develop a way of evaluating the present. But I am only using expectation; I have not actually experienced the future as one does in the fantasy of preknowledge.

We recognize that the notion of refracted time is based upon a spatial conceptualization of time. In feeling guilty about an experience that one only anticipates but has not yet experienced, one is not perceiving time in terms of the moment-by-moment sensations of the present that one is directly experiencing. That is, one is not feeling guilty about an event that is taking place at precisely this instant. So, while the notion of refracted time develops from highly subjective perceptions, it does represent a feeling about time that many of us experience.

Now, keeping in mind our notion that on the Experiential Inventory future-oriented people may be involved more with their personal and historical futures than past- and present-oriented people, we discover that there are no statistically significant relationships between experiential orientations to time and the fantasy of knowing the future.[21] Where in time persons place their most important life experiences does not relate to their willingness or unwillingness to fantasize about the personal or historical future.

THE TIME OF EXCHANGE AND SHARING

All of this brings us again to a serious limitation of the Money Game: its predication on fantasy. People are not actually buying time in the Money Game. The Money Game does not provide a wholly logical task because we do not ask respondents to sacrifice anything to buy units of time.[22] We merely ask whether they would *like* to fantasize about recovering the past and knowing the future.

Despite its limitations, however, the Money Game has provided us with important information. Essentially, its purpose is to measure the degree to which people are willing to fantasize about recovering the past and knowing the future. Although there were many who did not wish to fantasize about recovering the past and knowing the future, some did want to; and men and women did disagree on how to use their fantasies. Posttest interviews revealed that many men had difficulty playing with time in the way the Money Game demanded. Most believed that time acquired through fantasy should have some useful value; hence, they worked their fantasies into real life activities.

Perhaps the most interesting finding from our posttest interview responses was that our students changed tenses as they thought about time in realistic and fantasy terms. "If I could have a year back," one young man said, "it would be my senior year of high school. I would work hard and get into a good college. I screwed up too much!" In this example, the sequence of tenses goes from the conditional to the present and finally to the past. In only a moment the fantasy of recovering past time was turned into a fantasy of altering the present. Women, through similar posttest interviews, revealed that their common reason for recovering past time was to relive it. "I could be nine again and go to the zoo on Sundays with my father," one young woman reminisced. "It was nice." Here, the conditional tense moves into the past tense, but never does her awareness of present reality destroy the wish-fulfilling reverie. Women seem to accept the limitations of pretending and are willing to separate the realms of fantasy and reality. In contrast, men shift reality around so that the time they acquire through fantasy will be able to affect real ongoing situations.

One final observation on the Money Game. That the corpsmen and corpswomen were being misled by the instructions in systematically different fashion was apparent from the time we began testing. The major problem concerned the economics of the game. If one purchased a year from the past, men wondered if that meant that one's future life is shortened by the same amount of time that one has recovered from the past. Apparently, men were confused by the idea of getting something for nothing. In contrast, women's confusion was best expressed by the question: If I buy back a year from my past, may I also buy a year of my future?

Men were confused because they felt some natural sense of *exchange* had been violated. Deprived of the opportunity to relinquish a commodity in exchange for time, they found the familiar market rule negated, and money itself became valueless. Women, however, seemed more concerned with accumulated experiences and with how to share them. They asked, "How much can be taken, and will there be enough to give later on?" The matter of exchange and sharing is discussed fully in Chapter 11; here we should note that men, in thinking about exchange, have been influenced by their understanding of the rules of the market. When these rules are broken, men no longer seem certain how to deal with money. Women, taught to be less concerned about financial matters and marketplace relationships and more concerned with their feelings, wonder instead how to share their emotions with other people. Thus, we might say that *socioeconomic* matters influence men's performance in the Money Game while *social emotional* matters influence women's performance.

CONCLUSION

Despite its fantasy nature, the Money Game permits an exploration of several new perceptions of time. We can begin to explore inner feelings about the attitudes toward time that are as important as intellectual conceptualizations and measureable perceptions of the extension of time zones. The Experiential Inventory allows a look at the placement of our experiences in time; the Money Game allows a look at our hoping and wishing about time. The data thus far supports the viewpoint, expressed in Chapter 1, that our perceptions of any one time zone need not relate to one another nor even be consonant. Indeed, one senses that the attitudes we hold toward a time zone may be in conflict.

NOTES

1. Cf. Henri Bergson, *Time and Free Will* (London: Allen and Unwin, 1910); and *Duration and Simultaneity, with Reference to Einstein's Theory*. Tr. by Leon Jacobson (Indianapolis: Bobbs-Merrill, 1965).
2. Herbert Fingarette, *The Self in Transformation* (New York: Basic Books, 1963). p. 207.
3. William Henry, *The Analysis of Fantasy* (New York: Wiley, 1956), p. 1.
4. See Philip Merlan, Time Consciousness in Husserl and Heidegger, *Philosophy and Phenomenological Research*, 8 (September 1947).
   Medard Boss wrote about these notions of authentic and inauthentic fantasies in terms of the re- or preexperiencing of original time or life space time and the imagining of indirectly encountered history: "Original time is no external framework consisting of an endless sequence of nows on which man can eventually hang up and put into proper order his experiences and the events of his life. . . . Man's original temporality always refers to his disclosing and taking care of something. Such original temporality is dated at all times by his meaningful interactions with, his relating to, that which he encounters. Every now is primarily a now as the door bangs, a now as the book is missing, or a now when this or that has to be done. The same holds true for every then. Originally, a

then is a then when I met my friend sometime in the past or a then when I shall go to the university again. Medard Boss, *Psychoanalysis and Daseinanalysis.* Tr. by L. B. Lefebre (New York: Basic Books, 1963), p. 45.

5. Leonard W. Doob, *Becoming More Civilized: A Psychological Exploration* (New Haven: Yale University Press, 1960).

6. The introduction of money raised an important issue, the law of diminishing utility. There is a difference between asking people the financial value of a year of their personal past and an hour of their personal past.

7. Joost A.M. Meerlo, *The Two Faces of Man: Two Studies on the Sense of Time and on Ambivalence* (New York: International Universities Press, 1954), p. 45.

8. *Ibid.,* pp. 87–88.

9. *Ibid.,* p. 96.

10. To compute mean scores, the financial values listed as *a* to *e* on the Money Game were transposed to the numbers 1 to 5.

11. On this point, see Bertrand Russell, On the Experience of Time. *Monist,* 25 (1915): 212–33.

12. Karl Mannheim, *Essays on Sociology and Social Psychology* (New York: Oxford University Press, 1953), p. 111.

13. Edward Tiryakian, The Existential Self and the Person. In *The Self in Social Interaction,* C. Gordon and K. J. Gergen (Eds.) (New York: Wiley, 1968), p. 13.

14. Maurice Merleau-Ponty, *Phenomenology of Perception* (New York: Humanities, 1962).

15. See Eric Voegelin, Immortality: Experience and Symbol. *Harvard Theological Review,* (1967): 60.

16. Martin Heidegger, *Being and Time* (New York: Harper & Row, 1962), par. 65.

17. Cited in Joost Meerloo, The Time Sense in Psychiatry. In *The Voices of Time,* J. T. Fraser (Ed.) (New York: Braziller, 1966) p. 240.
    According to Mircea Eliade, the historical past and future are said to be the time of the *sacred,* in contrast to the personal past and future which are said to be the time of the *profane.* See Eliade, *The Sacred and the Profane.* (New York: Harper & Row, 1961).

18. Quoted in Miguel de Unamuno, *Tragic Sense of Life.* Tr. by J. E. Crawford Flitch (New York: Dover, 1954), pp. 211–223.

19. On this point, see Ernest Becker, *The Denial of Death* (New York: Free Press, 1974).

20. On this point, see Erik H. Erikson, *Childhood and Society* (New York: Norton, 1950); and *Insight and Responsibility* (New York: Norton, 1964).

21. The values are found in Table 18.

22. A better procedure might have been to insist that the students make a real trade for these time allotments, in the manner that Doob described. See Doob, *op. cit.*

CHAPTER SIX

FUTURE ORIENTATIONS
AND AVOIDANCE:

THE AFFECT OF SOCIAL
ROLES
ON PERCEPTIONS OF TIME

Philosophical and psychological theorists on time perception can be placed in groups, and each group considers a particular time zone to be more important than other zones. Psychologists Rollo May and George Kelly consider the future to be dominant, primarily because they believe that human development is future oriented, and of all human activities, intending, planning, and expecting are most significant. In Ludwig Binswanger's words, the "primary phenomenon of the original and authentic temporality is the future, and this future in turn is the primary meaning of existentiality."[1]

If one examines psychological research on human development and the nature of the life cycle,[2] one finds an emphasis on the future-directedness of human behavior. Based on his own experiments, Stephen Klineberg observed that as one passes from preadolescence to adolescence, his or her perception of the future undergoes significant and measurable change.[3] According to Klineberg, people learn to recognize links between the present and future and in this way acquire a more realistic idea of the future. As young people mature, they learn that they can control certain future outcomes.[4] They learn the meaning of intention and possibility.

But inexorably, we move toward the future, aging a bit more in every moment. To believe that by moving toward the past (as if one could), one can postpone or avoid the future is futile. Doing this only establishes a new future course. To deny the future is to deny existence.[5]

Whether one prefers a linear or spatial conception of time, the inevitability of the future cannot be disregarded. Although we agree we cannot know what the future will bring, we can still have expectations and make preparations. In fact, some people are so preoccupied with their own expectations and intentions, they may disregard the present. Other people may think very little about the future, preferring to take each day as it comes. Thus, the distinction between having and not having expectations is really the difference between living *for* the moment and living the moment *in preparation for* the future. It is also a distinction, as Florence Kluckhohn proposed, between being and becoming.[6] An expectation is not merely another passing thought about the future; it is the antecedent of deliberation and action. George Kelly emphasized how important planning for the future is, stressing the risks they face by not doing so. The psychologist Gordon Allport expressed the idea most powerfully: "The most important question we can ask about any mortal is what does he intend for the future."[7] Not only does the future take precedence because human activity can only move forward;[8] equally important, human activity is action-oriented. Moreover, to act effectively—that is, to approach the future realistically—one must have intentions and expectations.

The word *realistically* is important. In our review of linear and spatial conceptions of time, we stressed that the future's content cannot be known before the future arrives, even though some people may fantasize that they know. But fantasizing is not a realistic approach to the future. Instead of pretending to know what the future holds, one can plan and work out his or her expectations and intentions realistically, realistically because one does not disregard the inevitable passage of time; instead, one uses the present to prepare for the future.

In addition to emphasizing future-directed activity, Allport uses the word *mortal*. One presumes that he selected this word because it implies the finite quality of existence and because to be involved with one's own future, either by expecting or intending, is to be cognizant of dying. In the economics of existence, time is the ultimate scarcity. To face the future—indeed, to challenge it with expectations and intentions—is one way of coping with some of the future's uncertainties. Whatever our expectations or intentions, we are aware that our lives last only so long. We can never be certain we will live long enough to see our plans realized.

Given Allport's statement, how one chooses to define the future chronologically or how one visualizes it metaphorically—for example, "a galloping horseman," "a bird in flight," or "a quiet motionless ocean."—is inconsequential.[9] What mattered to Allport is that in contemplating and dealing with the future, intending is the only realistic strategy. Other strategies, such as predicting the future or fantasizing that one knows its contents are unsatisfactory because they imply changing the very nature of time. If one assumed a purely linear conception, the only way to deal with the future would be to sit back, wait patiently, and let it happen. In the context of the linear succession of instants, the future's arrival can neither be hastened nor postponed, just as past moments cannot be retrieved or relived. One gets older and watches the chronological past grow, the chronological future diminish.

If one assumes a spatial conception,[10] however, the best way to approach the future, as Allport indicated, is to fill the present with expectations and intentions. Here, expectations represent a way of preparing for the future, for they make future outcomes seem more certain and believable.[11] In a spatial conception of time, moments no longer seem to pass in succession, making us believe that our expectations do, in fact, produce desired future outcomes. A passage from George Herbert Mead captures this complex sense of subjective time perception:

A present, then, as contrasted with the abstraction of mere passage, is not a piece cut out anywhere from the temporal dimension of uniformly passing reality. Its chief reference is to the emergent event, that is, to the occurrence of something which is more than the processes that have led up to it and which by its change, continuance, or disappearance, adds to later passages a content they would not otherwise have possessed.[12]

The point of again mentioning the spatial conception of time is to assert that human beings *must believe* that they *can* in some way affect future outcomes. If one believed *only* in the linear conception of time, one would do little but wait for events to happen. But we know from experience that people do not idly await events; they actively try to shape them by the mechanisms of expectation and intention. Thus, while the linear view of time is more appropriate as a metaphor for chronological time, the spatial view is more appropriate existentially. It comes closer to describing the way we actually *experience* time in the course of our daily lives. As Siegfried Kracauer wrote: "Much as the concept of it is indispensable for science, it [linearity] does not apply to human affairs."[13] In spatial conceptions like Mead's, expectations and intentions become a special kind of present activity leading to their own kind of future result: Men and women begin to believe that they have played a role in causing certain future events to happen.[14]

THE STUDY

Our task in this chapter is two-fold. First, we examine how we expect or anticipate the future and how our expectations and anticipations make planning and action possible.[15] Second, we examine social role, a factor outside ourselves that influences our expectations. In effect, we are exploring patterns of socialization, the ways in which people learn about time (and other things) and the ways in which their overall attitudes toward the future are shaped.[16]

As we examine expectations and this "realistic" approach to the future, we may be tempted to believe that people at all times have carefully conceived expectations. We cannot be certain of this, however. If we ask people to list their expectations, they may become defensive, feeling unwilling to divulge things they either don't want to divulge or which they consider to be irrelevant to their lives. Other people, when asked to contemplate the temporal horizon, may report that the future is simply not worth speculating about. As the philosopher Robert Brumbaugh wrote: "Individual persons may differ drastically in their actual time location. This is an actual fact, not a mere possibility; and some of its implications are being discussed now in psychology."[17] Although the act of expectation represents a functional adaptation to time, we must inquire whether it really *is* universal.

We may begin this inquiry by reexamining the results of the Experiential Inventory. The Experiential Inventory provides a list of 10 important life experiences, experiences that one has already had, is presently having, and expects to have in the future. The possible absence of expectations on the Experiential Inventory among some respondents does not imply that these people do not think about the future. According to Allport, expectation is not the only way to deal with the future. Others are to fantasize preknowledge and to try to predict future outcomes. Do people who use the realistic means (list their expectations) engage as well in unrealistic approaches to the future? Do they also fantasize preknowledge? Do they try to predict the future as much as those who do not list expectations, those who may deal *less* realistically with the future?

Before examining the willingness to predict the future, the distinction between the fantasy of preknowing the future and predicting it must be clarified. In fantasizing preknowledge we express a wish to know what the future holds, simultaneously recognizing that no one can possibly tell us. In predicting future events we are sufficiently confident that our knowledge of past and present circumstances allows us to make inferences about future events. We may feel that prediction involves risk, for no one can be certain of the future, but we are willing to accept the risk. One might argue that if to predict the future one needed to perceive a causal relationship between present and future, few would perceive it better than future-oriented people. For them, prediction would not be an unrealistic way to grapple with the future; rather, it shows that they believe they are able to grapple with it effectively—that is, to *alter* it. Milton Rokeach wrote about such a capacity:

Another clue to a future oriented time perspective is the belief that one knows or understands the future. The person believing this often accuses others of not understanding. Such a person, guided by his belief–disbelief, typically expresses overtly a greater confidence of what the future holds in store and a greater readiness to make predictions about the future.[18]

We cannot be certain, therefore, whether having expectations encourages or discourages one to make predictions about the future. What we can assume—at least a bit more reliably—is that people actively engaged in expecting and intending will perceive the future as being more important than those who do not engage in expecting.[19] Similarly, we could say that those engaged in expecting would confer *less* significance on the *present* than those *not* engaged in it.[20] We base this reasoning on the notion that to be involved with the future is to see the present as merely a time to plan. In that sense the future becomes dominant, the present recessive. Or at least future-oriented people see it that way.

THE TESTS

Two of the tests for studying time perception, the Experiential Inventory and Money Game, are already familiar to us. The other two tests discussed in the chapter are the Semantic Differential, a way of measuring people's perceptions of the *potency* of the past, present, and the future; and the Future Commitment Scale, a way to measure people's willingness to predict future outcomes.

The Experiential Inventory.   In our discussion of the Experiential Inventory in Chapter 4 we described a procedure for dividing respondents into groups with high, medium, and low mean scores.[21] To examine expectations of the future we recategorize our respondents into three different groups. In the first group are 31% of the men and 25% of the women; they listed no experiences in the near or distant future; and all their experiences were placed in the past or present. We called the group *future-avoidant*. The second group, those who list only one or two distant- or near-future experiences, represents 33% of the men and 41% of the women we studied. We call it the *middle* group. Those who listed three or more near- or distant-future experiences, represent the remainder of our respondents: 36% of the men and 34% of the women. We call this group *future-oriented*.[22]

The Money Game.   In our discussion of the Money-Game we examined the fantasy of knowing one's personal future. The responses we received were distributed on a scale of 1 to 5, with those showing greater willingness to offer more money to purchase more time from the past and future ranked highest. We also saw that some students paid nothing for any future time offered them. In this chapter our focus is not on the *degree of interest* in acquiring knowledge of the future as measured by the purchase price, but on the *willingness* of people to pay any amount of money, no matter how small, for the chance to know their personal future. In Chapter 5 we called people *fantasizers* if they were willing to pay any money at all for the chance to know the future. Those who refused to pay anything, even the smallest amount of money ($10) to know even the smallest unit of time (one hour) in the future were called *realists*.

The Semantic Differential.   Implicit in the perceptions of time are certain attitudes. People may place a variety of their experiences in the future, but that doesn't tell us what they *think* about the future, their attitudes toward it. Similarly, although one may perceive the present's duration as lasting a second or a year, what does this tell us about how he or she *regards* the present? Is one strongly influenced by the present? Is the present, in other words, "potent"? Is the present a pleasant time to live in; that is, does one regard the present as "good"?

We examine attitudes toward the past, present, and future and the concept of time by using the Semantic Differential Instrument developed by Charles Osgood and his associates.[23] We include the concept of time in order to examine whether some people may express their attitudes about this more general concept rather than about a particular time zone.

The Semantic Differential instructs respondents to express their attitudes toward various concepts or objects by using adjectives to describe these concepts or objects. The adjectives are presented in pairs of opposites—for example, good and bad. Furthermore, between each pair of opposites are seven points of gradation. Thus, a person may describe the future as being good, bad, or some measurable degree of goodness or badness in between. If a person feels that the future is neither good nor bad but essentially neutral, he or she can indicate this by "scoring" it 4.

Extensive research has found repeatedly that the Semantic Differential reveals two groups of attitudes: *evaluative*, which refers generally to the goodness or badness of the concept or object evaluated, and *potency*, which refers to the perceived power of the concept or object. Respondents in our research were given an evaluation and potency score for each of the three time zones, past, present and future, and for the concept time. Table 13 shows the Semantic Differential exactly as it was presented to respondents, with the groups and their respective scales identified.[24]

The adjective pairs that describe the potency group are: hard–soft, large–small, strong–weak, brave–cowardly, and rugged–delicate, the first word signifying greatest potency. Adjective pairs describing the evaluative group include clean–dirty, good–bad, kind–cruel, fair–unfair, and pleasant–unpleasant. Here, the positive evaluative word is the first.

One might say that the Semantic Differential confuses respondents as they attempt to express their attitudes toward the past, present, and future and time. Do respondents, for example, find adjective pairs like hard-soft, rugged–delicate, fair–unfair, helpful–unhelpful, appropriate or inappropriate in expressing their attitudes?

The results of the Semantic Differential, shown in Table 14, reveal that the adjectives provided might not have helped respondents in determining their attitudes toward the concept time, because the evaluation and potency values obtained clustered around 4.0, a neutral position. For the most part, however, the potency and evaluation scores for the past, present, and future fell toward the higher end of the scale, which means that, on the average, the past, present, and future were judged to be moderately good and moderately potent.

The correlations among potency and evaluation scores for the four concepts, shown in Table 15, reveal an interesting pattern. For women, there are very few correlations that do not reach a statistically significant level: The variables are strongly related to one another. In contrast, men revealed several instances where the correlations between variables were very low—for example, between the judged potency of time and the evaluation of the future ($r = .02$).

One relationship shown in Table 15 is important precisely *because* it fails to reveal a statistically significant level for either sex. This is the correlation between present potency and present evaluation (male $r = .17$; female, $r = .02$). Although the mean scores of these two variables were almost identical for men and women, the variables remained almost perfectly independent of one another. Apparently, how our respondents perceived the present's potency had nothing or little to do with how they perceived its goodness or badness.

The Future Commitment Scale. The Future Commitment Scale was designed to measure the degree to which people are willing to predict the future, or to accept events they may or may not be able to control. The Future Commitment Scale was based in part on an instrument designed by Alan H. Roberts and Robert S. Hermann.[25] Their scale, called a Certainty–Uncertainty Scale, instructed respondents to make a prediction about something and then indicate on a scale of 1 to 7 how certain they were about that prediction. An example might be: Predict the next President of the United States.

Before beginning their study, Roberts and Hermann hypothesized that dogmatic people would be more willing to make predictions than undogmatic people.[26] They also hypothesized that dogmatic people would be more certain about their predictions. In a sense these hypotheses define the dogmatic person as someone who feels he or she has special understanding of the future's content.[27]

The authors also suggested an alternative hypothesis: While dogmatic and undogmatic people may not differ in the degree of certainty about their predictions, dogmatic persons would reveal a greater variability between the certainty and uncertainty of their predictions than undogmatic persons. In fact, the results of the study confirmed this second hypothesis.

The Future Commitment Scale was designed with this variability of prediction in mind. The scale presents 31 statements concerning future occurrences—for example, "Some day the entire world will be run by the United Nations," and "Man someday will find life on many planets." So-called global statements like these were combined with more personal statements such as, "I will have a good and successful life." Our respondents were instructed to agree or disagree with each statement or register their unwillingness to make a prediction by choosing the option, "I can't say." The number of "can't say's" measures one's future commitment. The *more* "can't say's," the *less* a person is willing to make predictions about the future, and the less he or she is committed to it.

The Future Commitment Scale and the frequency distribution for the three response options is presented in Table 16.[28] Although the table indicates that the students were less willing to commit themselves to issues about their own lives than to global issues,[29] we also want to know whether they were willing to predict certain global items more than other global items. To do this, a factor analysis was performed on all 31 items.[30] Results of the factor analysis shown in Table 17 indicate that not only have respondents differentiated global items from personal items, they also have differentiated *among* global items, putting them into three groups.

The first of these groups was labeled Immediate Social Commitment, because the items comprising it dealt with contemporary social issues. Slightly pessimistic in tone, the second group of items refers to distant events, both in time and space, and accordingly was labeled Distant Cosmic. Finally, the more hopeful items of the third group lie in time and space somewhere between those of groups one and two. Thus it was called Mediate Global. In addition, a group of items refers to personal events. These seven items comprise the Personal Commitment Scale (PCS).

In all, the Future Commitment Scale yields five results: total future commitment score for all 31 items and four individual group or subscale scores, including the Immediate Social Commitment Scale, the Distant Cosmic Commitment Scale, the Mediate Global Commitment Scale, and the Personal Commitment Scale. In this and later chapters, however, we discuss only the total Future Commitment Scale and the Personal Commitment Scale.

If one examines the individual mean scores of these scales (Table 18), one notes that men consistently reveal greater willingness to predict than women,[31] but only about global items. Both sexes began to select the "can't say" option more frequently for personal items.

Results. Reactions of future avoidance.   Earlier in this chapter we offered the hypothesis that *future-oriented* people—people with three or more future experiences on the Experiential Inventory—would be less willing to make predictions about the future than *future-avoidant* people—those listing no near-future or distant-future experiences on the Experiential Inventory. This hypothesis was based on the assumption that having expectations represented a realistic approach to the future, whereas predicting the future represented an unrealistic approach. Thus, we could interpret the act of predicting the future as a compensation for the absence of expectations. We also offered an alternative hypothesis: Having a great many expectations indicates that one has a sense of being able to affect outcomes. Future-oriented people, therefore, should predict more than future-avoidant people.

The results shown in Table 9 partially support the first hypothesis. Only future-avoidant men engaged significantly in predicting their personal future ($\chi^2 = 4.91$, p $> .05$).[32] This finding is especially interesting, given the fact that predicting global items on the Future Commitment Scale is unrelated to the three experiential time orientations $\chi^2 = 2.00$ for men; $\chi^2 = 2.64$ for women), even though personal and global prediction are themselves highly correlated ($r = .63$, $p < .01$ for men; $r = .60$, $p < .01$ for women).[33]

If we may now say that future-avoidant men compensate for their lack of expectations on the Experiential Inventory by making predictions about themselves, future-avoidant women compensate for their lack of expectations by fantasizing knowledge of the personal future on the Money Game ($\chi^2 = 5.88$, $p > .05$). This finding is also shown in Table 19, where fantasizers represent people who offered at least $10 to buy any unit of time of their personal futures. Let us also remember that about 47% of the corpsmen and corpswomen refused to pay anything for the chance to know their personal futures.

Not only do these data suggest that men and women have different ways of dealing with the future in the absence of expectations, they also increase our understanding of the ways themselves. In predicting, people are expressing a certainty about the future to compensate for having made no preparation *for* the future. Why prepare if one is already certain of the outcomes? Or, in the case of women who list *no* future expectations, one can *wish* to know the future and thus be untroubled by one's lack of preparedness. One can always try to solve very real problems through fantasy.

Time potency and present evaluation.   At first glance, the potency and evaluation mean scores shown in Table 14 might suggest that our respondents were unable to use the adjectives provided them in the Semantic

Differential to express their attitudes about the past, present, future and the concept time. The mean scores, after all, hover about the neutral points on each scale, the points between the paired adjectives.[34] But examination of experiential orientations revealed on the Experiential Inventory and comparison of them with similar orientations revealed on the Semantic Differential indicates that the respondents were indeed expressing strong attitudes, at least on the Semantic Differential, and not choosing the neutral point at all.

For example, although both the potency and evaluation means for the concept time were seen to hover near the neutral point (Table 14: potency $\overline{X}$ 4.8, evaluation $\overline{X}$ = 4.6), time's potency was rated significantly higher by future-oriented men than by future-avoidant men (χ = 7.56, $p < .05$).[35] However, future-oriented men evaluated the present lower (below the combined male–female evaluation mean) than future-avoidant men (χ^2 = 6.05, $p < .05$). And, although the individual Semantic Differential mean scores did not distinguish one sex from the other (Table 14), there was no association between Semantic Differential values and experiential time orientations for women (Table 19). Because of their expectations for the future men tend to devalue the present, or at least view it merely as a time of preparation.

Prediction and the Perceived Extensions of Time Zones. The concept of future prediction is a complicated one, raising fundamental questions. Is commitment the result of feeling that one will actively take part in creating the outcome of an event, or that one will live long enough just to know the outcome?[36] Or, does commitment imply that because persons will not live to experience the events they have predicted, they need have no responsibility for how these predictions turn out? A respondent might reason, "I might just as well predict; no one alive now will know whether I am right or wrong." Or he may reason, "I won't live long enough to find out whether I'm right or wrong so why *not* predict?" A third approach might be, "Because I will not be around, I had better not say anything because I do not want to be held responsible for something whose outcome I cannot even know." Prediction, then, might go hand in hand with the idea that if one feels one will live long enough or if the future will extend itself far enough, the prediction may come true. In other words, if one lives long enough, one is likely to see or know just about anything.

At first glance we might miss another aspect of prediction: How a person defines chronologically the period of the future that he or she is willing or unwilling to predict. In the simplest terms, some people might attempt to predict the future because for them the future begins seconds from the present. Other people, for whom the future is years distant, may be less willing to make such an attempt.

To examine the degree of association between people's willingness to predict the future and the way they perceive it in terms of its starting and ending points, we examine the relationships between respondents' performance on the Future Commitment Scale and the Duration Inventory.

The correlations between the Future Commitment Scale and the six Duration Items shown in Table 20 reveal no statistically significant correlations for men, but they do show several significant correlations for women. Although the correlations are of a low magnitude, willingness to predict the future increases as a function of extending the distant past ($r = -.24, p < .05$), the near future ($r = -28, p < .05$) and the distant future ($r = -.26, p < .05$).[37] The correlations suggest therefore, that the farther away the past and the future seem to lie, the more willing women are to make predictions about the future. These correlations also suggest that women are more willing to make predictions when the events to be predicted are located in the distant future.

INSTRUMENTALITY AND EXPRESSIVITY: THE EFFECT OF SEX ROLES ON
PERCEPTIONS OF TIME

The most important implications that may be drawn from the relationships we now have uncovered is that men and women view the future differently. Men are constantly attempting to construct their own futures, believing that rewards for their present efforts will come not in the present, but in the future. Said differently, men are socialized not to expect immediate gratifications of their efforts. This form of socialization was described by Talcott Parsons in his concept of the *instrumental interaction*, an interaction between people in which neither person expects immediate gratification.[38] Instead, both persons use the interaction as an *instrument* to achieve later goals and later gratifications.[39] The present gains importance because mainly during this period people prepare for the future. Thus, one may consider the present as a period of potential rather than realization, for present plans and expectations cannot be assessed, much less experienced, until the future is assessed and experienced, or until they can be assessed retrospectively from the vantage point of the future. Only in the future can the unfamiliar be mastered and the real goals of life be attained.

Normally characterized as a masculine form of interaction, the instrumental interaction implies, according to Parsons, that the outcome of the interaction can never be certain. Because the ultimate rewards of the interaction lie in the future, participants in an instrumental interaction expect that no outcome of the interaction can be predicted with certainty. Perhaps the best way to understand what Parsons meant by an instrumental interaction is to refer to the familiar notion of means and ends. People are preoccupied with the *means* to some future end or outcome, but the end or outcome can only be experienced later. As John Dewey wrote, "Structure is constancy of means, of things used for consequences, not of things taken by themselves or absolutely."[40]

With this in mind, one might reason that by its very nature, the instrumental interaction *discourages* prediction of the future, if only because it denies predictive certainty. It would also discourage the fantasy of knowing the future, because it implies that not only must one learn to postpone gratifications, but one must also live with the doubt that gratifications ever will be realized.[41] In the way Parsons describes the instrumental interaction, the only approach to the future is expectation. Moreover, expectations become especially important to people involved in instrumental interactions because existence is finite. To have expectations would be fruitless if time had no personal limits or if future events could be known. Expectations, therefore, may be likened to fragments of one's self placed forward in time, which are then integrated into one's life through work and the passage of time.

If Parsons' instrumental interaction, with its emphasis on future gratifications and its *de*emphasis of the present, is characteristically masculine, the *expressive interaction* is characteristically *feminine*. Emphasized here is immediate and direct gratification. Rather than shaping future rewards, the present provides immediate rewards, and the exchange between people becomes an end in itself. If the instrumental interaction encourages preparation for the future and thus makes the present a zone of *potential*, the expressive interaction encourages acceptance of the present as a zone of *being*. Because gratifications are immediate in the expressive interaction, people engaged in such interactions would fantasize about knowing the future. In the expressive interaction one lives in the present.

The future, with all its uncertainties, can only come later; as one reaches each future stage, one is actually experiencing a new present.

The concept of an expressive interaction may explain some of the results we have examined in this chapter. By being socialized in this interaction, women may be encouraged to visualize their expectations as occurring in a series of interrupted presents and as ends in themselves, rather than as means toward some future end which they create in the present. As Anne Hatcher observed: "Each moment has the weight of eternity behind it and hence achieves a permanence beyond its individual passing. Each pleasure is something greater and more lasting than itself."[42] Not only is the temporal character of women's expectations different from the temporal character of men's expectations, but women, socialized to play a different role from men, learn also to accept a different concept of the future. Indeed, Allport's original pronouncement now seems more applicable to men than to women: "The most important question we can ask about any mortal is what does *he* [emphasis added] intend for the future."[43]

These last notions assume added significance when we discover that future-*oriented* women scored significantly higher on intelligence tests than future-*avoidant* women ($\chi^2 = 8.62, p > .01$).[44] This finding, however, does not hold for men ($\chi^2 = 2.39$, n.s.) of equal measured intelligence. Women, therefore, may need to have "more" intelligence than men to have expectations about the future. They may have been socialized to engage in expressive interactions and thus to expect immediate gratification. Regardless of their intelligence, most men are socialized into instrumental interactions, thus to expect future gratifications. For them, the present is a time of preparation; for women, it is a time of having.

That a future orientation should be associated with high intelligence scores for women but not for men is an interesting finding when we recall that Wechsler's classical definition of intelligence, meant for both sexes, stressed the importance of having expectations for the future: "Intelligence, operationally defined, is the aggregate or global capacity of the individual to act purposively, to think rationally and to deal effectively with his environment."[45] Although our own tests may contain flaws and our sample may not be fully adequate, they do show that measured intelligence is not significantly related to a future orientation on the Experiential Inventory for men in the same way as for women. In addition, intelligence does not significantly correlate with the other two characteristics we have considered in this chapter—a willingness to predict the future (male,

$r = -.07$; female, $r = -.17$), which implies a confidence in one's abilities to make about the future; and fantasies of knowing it, in which no risk is involved (male, $r = -.06$; female, $r = -.04$).

CONCLUSION

On the basis of our findings and our discussion of the differences between the characteristically masculine instrumental interaction and the characteristically feminine expressive interaction, we might conclude that men experience a feeling of being disengaged from the present which women do not feel. We may also speculate that women tend to experience immediate gratification from their present efforts and work, whereas men consider expectations of future achievement to be more important than personal achievements experienced in the present because they fear that completed activity—what Abraham Maslow called self-actualization—may mean the death of activity.

More important than any specific sex difference or speculation that we may make at this point is that the data discussed in the past several chapters indicate that society, through the imposition of social roles, influences the way men and women perceive time. We cannot say that, given the way they use the present to prepare for the future, men consciously adopt the spatial conception of time, while women, because they accept the present as a time of having, consciously adopt a linear conception. We can say, however, that an external influence—the socialization of men and women into social roles[46] and the styles of interaction common to these roles—affects experiential orientations to time, the fantasies of recovering the past and knowing the future, the willingness to predict the future, and attitudes toward the past, present, and future.

In our sample persons who failed to list expectations on the Experiential Inventory apparently do not believe that marriage, parenthood, and career are significant experiences. At least they have not listed any of these experiences. Specifically, the findings show that men orient their expectations toward a style of interaction that Parsons calls instrumentality; thus, they ascribe a controlling or *agentic*[47] quality to their expectations. Women, however, seem to orient their expectations and perhaps even their perceptions of time itself toward a style of interaction that Parsons calls expressive; thus they ascribe a continuity to time. They achieve from this a *communion*[48] with both the present and future. Prediction of the future and the fantasy of preknowledge could represent respectively the agentic and communing forms of dealing with the future when one has not actively established specific expectations.

The instrumental–expressive distinction described by Parsons provides an important framework for a study of time perception. It suggests that women perceive time as a continuous flow, while men perceive it as discontinuous, fragmented. In the following two chapters we continue to explore what we have called spatial and linear conceptions of time. We also examine the possibility that women derive their concept of continous time from a unique view of it as *linear*, while men derive their concept of discontinuous time from an equally unique view of it as *spatial*.

The data we have inspected so far cannot confirm this, but we are beginning to sense the social and psychological disruptions young men must feel as they free themselves from their parents and start to experience the demands of instrumental interactions. One wonders whether these early disruptions eventuate in a sense of alienation and aloneness among adolescents and among many adults. Or perhaps these disruptions help stimulate young people to plan accomplishments for the future or to create continuity among past, present, and future.

Before we return to these questions and to investigations of spatial and linear conceptions of time, a word of caution is needed. The ways in which people perceive time are fascinating and seemingly endless, and any one of them could be researched extensively. We must assess our findings carefully, however, because as significant as they may be to researchers, they may not seem so to men and women in the course of their daily activities, or even in the course of a lifetime. We must not let our own interests betray us into thinking that time matters equally to all people, although all of us have life goals, emerging identities, and great interest in them. However, on the basis of the corpsmen and corpswomen's reactions to the various instruments, we do sense that the perceptions of time we are studying were *not* irrelevant or uninteresting. Indeed, many of the students found the instruments, in their words, "threatening," "frightening," and "intriguing." Their post-test reactions indicated that the various instruments were causing them to reflect on issues which they themselves called fascinating and important. Thus, while we will continue to be careful not to assume that our own interest in time perception perfectly coincides with our respondents' interest in it, we do have evidence that our research was touching upon very critical concerns in the lives of the students participating in this work.

NOTES

1. Ludwig Binswanger, The Case of Ellen West. In *Existence*, R. May, E. Angel, and H. F. Ellenberger (eds). New York: Basic Books, 1958, p. 302.

2. Cf. Stephen L. Klineberg, *Structure of the Psychological Future*. Unpublished thesis, Harvard University, 1966; and Melvin Wallace and Albert I. Rabin, Temporal Experience. *Psychological Bulletin*, 57 (1960): 213–236; see also Thomas J. Cottle and Stephen Klineberg, *The Present of Things Future*. (New York: The Free Press, 1974,) esp. Chapters 1, 6, and 7.

3. Stephen L. Klineberg, Changes in Outlook on the Future between Childhood and Adolescence. *Journal of Personality and Social Psychology*, 7, No. 2 (1967): 185–193; and *Structure of the Psychological Future, op. cit.*

4. See Cottle and Klineberg, *The Present of Things Future*. (New York: The Free Press, 1974) especially Chapter 3.

5. Gordon Allport, *Pattern and Growth in Personality* (New York: Holt, Rinehart and Winston, 1963).

6. Florence R. Kluckhohn, Dominant and Variant Value Orientations, in Kluckhohn and Fred L. Strodtbeck, (Eds.). *Variations in Value Orientations* (Evanston, Ill.: Row, Peterson, 1961).

7. Allport, *op. cit.* p. 223.

8. On this point, see Stephen L. Klineberg, Change in Outlook.

9. These metaphors are taken from a study by Robert H. Knapp and John T. Garbutt, Time Imagery and the Achievement Motive. *Journal of Personality*, 26 (1958): 423–434.

10. Cf. Henri Bergson, *Time and Free Will* (London: Allen and Unwin, 1910); and *Duration and Simultaneity*. Tr. by Leon Jacobson (Indianapolis: Bobbs-Merrill, 1965).

11. On this point, see R. B. Yaryan and L. Festinger, Preparatory Action and Belief in the Probable Occurrence of Future Events. *Journal of Abnormal and Social Psychology*, 63 (1961): 603–607; and Joel Cooper, Linda Eisenberg, John Robert, and Barbara S. Dohrenwend, The Effect of Experimenter Expectancy and Preparatory Effort on Belief in the Probable Occurrence of Future Events. *Journal of Social Psychology*, 71 (1967): 221–226.

12. *George H. Mead On Social Psychology*, Anselm Strauss (Ed. (Chicago: University of Chicago Press, 1964), p. 332.

13. Siegfried Kracauer, Time and History. In *History and the Concept of Time* (Middletown, Conn.: Wesleyan University Press, 1966), p. 69.

14. Cf. Friedrich Kummel, Time as Succession and the Problem of Duration. In *The Voices of Time*, J. T. Fraser (Ed.) (New York: Braziller, 1966).

15. On this point, see Talcott Parsons and Edward A. Shils (Eds.), *Toward a General Theory of Action* (New York: Harper & Row, 1962).

16. See Cottle and Klineberg, *The Present of Things Future*, esp. Chap. 6.

17. Robert Brumbaugh, Logic and Time. *Review of Metaphysics*, 18 (June 1965): 656.

18. Milton Rokeach, *The Open and Closed Mind*. (New York: Basic Books, 1960), pp. 52–63.

19. On this point, see Yaryan and Festinger, *op. cit.*

20. On a related issue, see Evelyn Glenn, Volition: A Study of Factors Related to Level of Experienced Volition. Unpublished thesis prospectus, Department of Social Relations, Harvard University, 1967.

21. Means were computed by adding the number of the time zone assigned to each experience and dividing by 10, the number of experiences.

22. For the rest of this chapter the term *future-oriented* refers not to a mean score on the Experiential Inventory, but to people who list at least three near- or distant-future experiences. On the basis of these new percentage distributions, neither sex can be considered more future-oriented than the other. Only an insignificant percentage difference between men and women appears in both the future-avoidant and future-oriented groups. The Experiential Inventory instructed people to list their expectations rather than intentions although experiences can essentially be interpreted as intentions.

23. Charles Osgood and George J. Suci, Factor Analysis of Meaning. *Journal of Experimental Psychology*, **50** (1955), 325–338; and Charles Osgood, George Suci, and Percy H. Tannenbaum, *The Measurement of Meaning* (Urbana: University of Illinois Press, 1957).

24. Adjectives were presented on seven-point scales, with 7 being recorded when the designated potency or evaluation adjectives was selected and 1 when its opposite was chosen. The middle option, 4, served as a neutral point. Final scores represent the means of the five adjective scales. Individual responses to the Semantic Differential were scored for more than 500 hospital corpsmen and corpswomen, not all of whom were included in the time perception research project. This additional testing was undertaken to corroborate the presence of Osgood's original potency and evaluation factors. Results of our own factor analysis clearly demonstrated the presence of the two factors.

25. Alan H. Roberts and Robert S. Hermann, Dogmatism Time Perspective and Anomie. *Journal of Individual Psychology*, **16** (1960): 67–72.

26. *Dogmatism* was defined in terms of a Dogmatism Scale originally designed by Milton Rokeach. See Rokeach, *op. cit.*

27. *Ibid.*

28. As was true with all other instruments, the Future Commitment Scale was pretested on several occasions with different populations in an effort to arrive at the most workable items.

29. Of special interest in Table 16 are the responses to items concerned with juvenile delinquency, racial problems, the cure of cancer and religious practices.

30. The analysis used a Varimax orthogonal rotation and principal axis solution. For the analysis, items were scored 2 if the "can't say" option was selected, 1 if either agreement or disagreement was recorded.

31. Low "can't say" scores mean a high willingness to predict the future, and vice versa.

32. For this computation, prediction categories were determined by dividing respondents into groups according to whether their scores fell above or below the combined male and female Personal Commitment Scale mean. Low scores indicate high commitment.

33. For this computation, prediction categories were determined by dividing respondents into groups according to whether their scores fell above or below the combined male and female total Future Commitment Scale mean.

34. Cf. R. H. Knapp and J. T. Garbutt, Variation in Time Descriptions and in Achievement. *Journal of Social Psychology*, **67** (1965): 269–272.

35. To be precise, the potency ratings of future-oriented men fell above the combined male and female potency mean significantly more often than the scores of future-avoidant men.

36. On a related point, see J. E. Teahan, Future Time Perspective, Optimism and Academic Achievement *Journal of Abnormal and Social Psychology*, **57** (1958): 379–380.

37. The correlations show negative values because high commitment to the future is operationalized as *low* can't say scores.

38. Parsons and Shils, *Toward A General Theory of Action*. (New York: Harper and Row, 1962.)

39. Cf. *ibid.;* Talcott Parsons and Robert F. Bales, *Family, Socialization and Interaction Process* (Glencoe, Ill.: Free Press, 1955); and T. Parsons, R. F. Bales, and E. Shils, *Working Papers in the Theory of Action*.

40. John Dewey, *Experience and Nature* (New York: Dover, 1958), p. 72.

41. On this point, see Walter Mischel, Preference for Delayed Reinforcement: An Experimental Study of a Cultural Observation. *Journal of Abnormal and Social Psychology*, **56**, (1958): 57–61.

42. Anne Hatcher, Universals in John Updike. Unpublished honors thesis, Department of Social Relations, Harvard University, 1967, p. 24.

43. Allport, *op. cit.*, p. 223.

44. Intelligence was measured with the General Classification Test (1945) and Mathematical Aptitude Test, which are administered to all entering military personnel. I.Q. is operationalized as the sum of these two scores whose intercorrelation is .74. To compute Chi Square Values, intelligence scores were divided at the combined male and female mean into high and low groups.

45. David Wechsler, *The Measurement and Appraisal of Adult Intelligence* (Baltimore: Williams and Wilkins, 1958), p. 7. Similarly, Lewin and Witkin have argued that increased mental acuity enhances one's capacity to tolerate and manipulate greater complexities of symbolic thought. According to these authors, intelligence makes it possible for one to depend less on the concrete guideposts of situational determinants. Both authors assert that intelligence leads to a greater awareness of reflections and expectations, hence, to an actual expansion of human experiencing. See Kurt Lewin, *Resolving Social Conflicts* (New York, Harper and Bros., 1948); and Herman A. Witkin, *Psychological Differentiation: Studies of Development* (New York: Wiley, 1962).

46. See Alex Inkeles, Social Structure and Socialization. In *Handbook of Socialization Theory and Research*, David A. Goslin (Ed.) (Chicago: Rand McNally, 1969); and E. Q. Campbell, Adolescent Socialization. In *Handbook of Socialization Theory and Method*, David A. Goslin (ed.) (Chicago: Rand McNally, 1969), pp. 821–859.

47. Bakan, *The Duality of Human Existence* (Chicago: Rand McNally, 1966.)

48. *Ibid.*

CHAPTER SEVEN

A SPATIAL CONCEPTION OF TIME:

THE CIRCLES TEST

Throughout the preceding chapters we have used two fundamental conceptualizations of time: the spatial and linear. The spatial conceptualization corresponds to the way we perceive time flow subjectively, how we feel about it; the linear conceptualization corresponds to the way we perceive it objectively, how we measure it. Men and women conceptualize time both subjectively and objectively, although because of their respective instrumental and expressive interactions, men characteristically may adopt the spatial conceptualization of time and women may adopt the linear conceptualization.

In this chapter we examine the spatial conceptualization of time. The central idea here is the relatedness of time zones. To clarify this concept of relatedness, we must remember the basic distinctions between the spatial and linear conceptualizations. We recall, for example, that in the linear conceptualization, moments are experienced as they occur, one by one. Past moments are irretrievable, and future moments can not be experienced until they become the present or until we reach them in the future. Moments are visualized as occuring in a line; hence, the term *linear*.

Although the linear perspective illustrates the way time flows *chronologically*, it does not necessarily describe the way people *feel* about time flow. The linear conceptualization derives from a measurable characteristic of time, not from a subjective feeling about it. As Friedrich Kummel wrote:

When time is considered as a succession of present moments, thus denying the reality of both past and future, then duration is excluded from temporality. If the past is thought of as being merely a "no-longer-present" and the future a "not-yet-present," the real basis of duration, that is, the presupposition that past and future are inherent in the present, is removed.[1]

Kummel's notion of the past and future as *inherent* in the present is quite different from the purely linear conceptualization, in which moments suc-

ceed other moments. This notion of Kummel's is what makes us conceptualize the passage of time in *spatial* terms. We recall Heidegger's idea
that the act of remembering links the past with the present, just as the acts
of expecting, intending, and hoping link the present with the future.[2]
Thus, in the spatial conceptualization of time flow, the past is never
lost—memory brings it back—and the future is not totally unknown
—through expectations and intentions it acquires substance.

As a result of conceptualizing time spatially, people begin to feel the
connections between—the relatedness of—past, present, and future in a
way denied them if they hold only to the linear conceptualization.[3] If we
perceive a relatedness between time zones, we find that the past and
future must be contained in the present, and we can use recollection to
retrieve the past and expectation to intimate the future. In fact, some
people would say that recalling and expecting make it seem as though the
past and future *reside* within the present.[4]

This spatial relatedness of time zones is consistent with certain notions
in recorded psychological research. David McClelland, for example, asserted: "It is as if the [achievement] need has served to relate present in
terms of a wider context . . . Motives seem to tie the present to the future,
the specific to the general and long run."[5]

A similar viewpoint was expressed by O. Hobart Mowrer in his definition of the concept of time binding.

Time binding, that is, the capacity to bring the past into the present as part of the
total nexus in which living organisms act and react, together with the capacity to
act in the light of the long-term future—is the essence of mind and personality
alike.[6]

We note that Mowrer ended with St. Augustine's words, "The present,
therefore, has several dimensions . . . the present of things past, the
present of things present, and the present of things future."[7]

McClelland's and Mowrer's notions (as well as others) give added cre-
dence to the spatial conceptualization. Both men are concerned with one
form of relatedness between time zones—time binding—a relatedness
alien to the linear conceptualization of time. But again, these are theories.
We have not yet demonstrated empirically that men and women do use
spatial conceptualizations to visualize time or to understand it. Nor do we
know the precise form such conceptualizations might take. Assume people
do conceptualize time spatially and that they do perceive relatedness of
past, present, and future. What form does this relatedness take? Does the
past reside *entirely* in the present or just in some part of it? Does the future

reside entirely in the present or just partially? Can we visualize three distinct and totally *unrelated* time zones, or one large time zone comprising all of the past, present, and future? Is there some form of relatedness between these extremes?

Another question is relevant to time zone relatedness. Does perceiving a relatedness between time zones have anything to do with perceiving one time zone as more important than the other zones? If we are to believe some of the philosophers and psychologists whose work we have reviewed, we would assume so.[8] The present would have to be most important, because only in the present do recalling and expecting take place. However, if we adopt a spatial conceptualization of time, we could consider the past or future to be equally or even more important. Although we know we can live only in the present, we can relate strongly to other zones, and in doing so, we give them significance.

In this chapter we explore the ways we perceive inter-zone relatedness and attempt to discover whether people ascribe greater significance to one zone or another. The instrument we used in our study is the Circles Test. Its objects were to encourage our respondents to think of time spatially, to provide them with a way of reporting their perception of the inter-zone relatedness, and to learn whether they perceived one zone to be more important than the other zones.

Major variables in the test were *temporal dominance* and *temporal relatedness*. Temporal dominance is the perceived significance of a particular time zone. If a person believes that the present zone is the most important, the perception is *present dominance*. Temporal relatedness is the perceived relatedness between time zones. We provided our respondents with a special format that encouraged them to think of time flow spatially. Our instructions for the Circles Test were:

Think of the past, present, and future as being in the shape of circles. Now arrange these circles in any way you want that best shows how you feel about the relationship of the past, the present, and the future. You may use circles of different size. When you have finished, label each circle to show which one is the past, which one the present, and which one the future.[9]

TEMPORAL RELATEDNESS

The first variable generated by the Circles Test was *temporal relatedness:* the degree to which the circles touched one another or overlapped either partially or completely. For example, if the past circle were completely

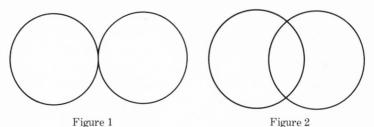

Figure 1   Figure 2

separated from the present circle, we scored the drawing 0 for relatedness. If the past circle touched the present circle at their respective peripheries (Figure 1), we scored it 2. If the past partially overlapped the present (Figure 2), we scored the drawing 4, and if the past circle appeared totally within the present circle (Figure 3), we scored it 6.

To determine the total relatedness score for a particular drawing, we added the individual relatedness scores of each pair of circles in the drawing. Thus, in Figure 4, the past and present overlap, earning 4 points, and the present and future overlap, earning 4 points. The total relatedness score is 8 points. The relatedness score, however, of an individual time zone cannot be determined merely by knowing the total relatedness score. Again in Figure 4, the past and future each scored 4 because of their association with the present. The present earns a score 8 because of its association with both past and future.

One numerical value may actually represent two discrete configurations. For example, each of the drawings shown in Figures 5 and 6 would be awarded 4 points. Figure 5 the overlapping between the past and present yields a score of 4. In Figure 6 the present touching both the past and future also yields a score of 4. But knowing that a relatedness score is 4 does not tell us much about the drawing. Accordingly, we must score the

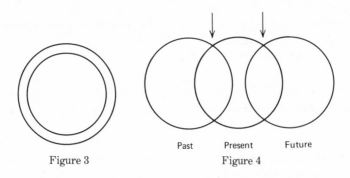

Figure 3   Figure 4

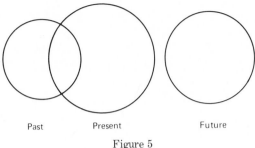

Past   Present   Future

Figure 5

total drawing as well as individual circles. Still, the important point is that the higher the score, the more related the time zones are to one another. Thus, we have at one extreme the configuration designated *temporal atomicity*, scored 0, in which the time zones are totally unrelated, and at the other extreme the configurations designated *temporal integration* and *projection*, scored 8 to 18, in which the time zones overlap partially or totally. Between these extremes is the configuration *temporal continuity*. In this, the zones *touch but do not overlap*.

Atomicity on the Circles Test is analogous to the spatial separation among time zones observed in responses to the Duration Inventory (Chapter 4). We recall that some respondents, for example, perceived the present ending minutes from now but the future commencing months from now. Blocks of time that were unaccounted for, therefore, were left empty between the conclusion of the present and the beginning of the future. The *continuity* configuration would then be analogous to the *linear* conceptualization—that is, time made up of consecutive moments. Finally, the *integrated* and *projected* configurations would be analogous to the *spatial* conceptualizations in which the time zones reside partially or

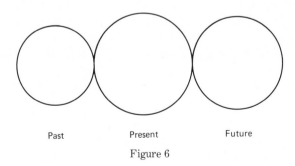

Past   Present   Future

Figure 6

totally within each other. For example, the configuration shown in Figure 4 represents, in Kummel's words, the past and future "inherent in the present," or in Heidegger's words, the future and past "turning in on the present."[10] (See Table 21 for frequency distribution of temporal relatedness scores.)

TEMPORAL DOMINANCE

For personal, social, and cultural reasons, some people perceive one time zone as being more significant than the other time zones. We have called this perception *temporal dominance*. If we examine the values associated with American culture, the future is dominant,[11] as it is in what Parsons called the *instrumental* interaction. Indeed, in any interaction where gratification must be postponed, the future is dominant. This contrasts with the dominance of the present in Parsons' *expressive* interaction. Naturally, some people may feel that the past is dominant. Indeed, we probably experience all sorts of pressures and values, each with its own emphasis on a particular time zone. For many people, the future is dominant merely because they are young. Yet, as Franz Kafka wrote, "The *space* of the future is often more than balanced by the *weight* of the past [italics added]."[12]

Temporal dominance on the Circles Test is ascertained by comparing the size of the circles in relation to each other. A circle receives 2 points if it is noticeably larger than another circle—that is, if we can assume it was drawn larger intentionally. For example, the configurations in Figures 5 and 6 would be scored this way: 0 points for the past circle, because it is *smaller* than both the present and future circles, and 4 points for the present circle, because it is *larger* than both the past and future circles. We score the future circle 2 points for although it is smaller than the present, it is larger than the past. Using this scoring system, we construct a scale in which the higher the score, the greater the perceived dominance of the time zone. We may also rank zones on the basis of their size. Thus, in Figure 5 the present would be ranked more dominant than the future and past, and in Figure 6 the future would be ranked more dominant than the past. One way of representing this ranking is: Present > Future > Past.[13]

Results Sixty percent of the people who took the Circles Test drew the atomistic configuration. The future was the most significant (or dominant) time zone, and the past was the least significant (Figure 7).[14]

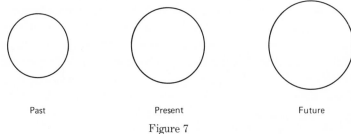

Past   Present   Future

Figure 7

These data reveal a perception of time in which the time zones are separated from one another. In this representation, moments do not succeed moments as they do in the linear representation of time flow. Indeed, the atomistic perception seems almost childlike in that a chunk of time called day is separated from the earlier and later chunks of day by something called night. Each day has its own historical boundaries. The expression "Tomorrow is another day," implying a new start to life, may be another way to verbalize the atomistic perception. In any event, people who view time atomistically can be called ahistorical: They deemphasize the past and its effect on the present and the future. In the frequency distribution of dominance scores shown in Table 22, the corpsmen and corpswomen performed almost alike, although women did show a slight tendency to make the present more dominant than the past.

The test seemed to stimulate respondents to visualize time spatially. The atomistic conceptualization, in which time zones are separate and discrete, is totally unlike the linear conceptualization of time. The linear conceptualization, in which moments succeed moments unendingly, carries the implication of *present* dominance, not the future dominance observed in the corpsmen and corpswomen. This raises the question: Is there some correlation between temporal relatedness and temporal dominance?

To answer this question, we computed a set of correlations between the temporal relatedness scores and the temporal dominance scores. The results of these correlations (Table 23) show that for men, temporal relatedness and dominance were not significantly associated. For women, however, the more dominant the future, the more atomistic the three time zones ($r = -.38, p < .01$).[15] Similarly, the more integrated the time zones, the less dominant the future. This is an important finding. If events occur continuously in time, if moment really does follow moment, and if the content of today does determine the content of tomorrow, as the linear conceptualization implies, the future, although the dominant zone on the Circles Test, tends to lose importance. At least this is so for women. In contrast, men revealed future dominance in both their atomistic and integrated drawings. But men who viewed time as continuous on the Circles Test most often drew the present and past circles larger than the future circle ($\chi^2 = 7.99$, $p < .10$; female $\chi^2 = 4.65$, $p = $ n.s.).[16] This finding is elaborated in Table 24. We may suggest, therefore, that women do not perceive of a plan as something that merely emerges from the natural flow of linear time; and unlike men, women may not fully believe that plans made in the present *can* affect future happenings.

From the correlations between temporal relatedness scores and temporal dominance scores, we note a relationship between temporal continuity—the form of relatedness most consistent with the linear perspective of time—and present dominance (see Table 25). We also find that if one perceives time in linear terms, one perceives the present as the "vital moment," as Bergson called it, or as a bridge linking the past to the future but having no significance apart from this. As one corpsman put it, "The present happens so quickly you almost don't see it. Just about the time we say 'now,' the moment has passed and we are confronted with the next 'now.' " Thus, the present in this linear conceptualization seems to contain only two of Augustine's dimensions, the "present of things past" and the "present of things future."

More generally, the correlations between the temporal relatedness and temporal dominance variables suggest that men systematically perceive the relative significance of time zones in a way that women do not. Apart from their perceptions of relatedness, men consider all three zones as a unit in deciding which zone is dominant; women seem to evaluate one zone at a time independently of the other two.

TEMPORAL DEVELOPMENT

One major difference was found between men's and women's test results. Perceiving the present as a fleeting and insignificant moment or as a bridge connecting the past and future is common among men, rare among women. This pattern, which we call *present insignificance* because the present is the least dominant time zone, was found too in the Experiential Inventory where 23% of the women and 42% of the men failed to list even one present experience (Table 1).

In contrast to this characteristically male perception of present insignificance, women revealed what we call *temporal development*—a movement of time forward from the past, through the present, and into the future; or backwards from the future to the past. Temporal development also suggests a perception of time "funneling"—expanding or contracting as one moves forward or backward through the three time zones. Time contracting as one moves forward toward the future, or what technically would be called *past-dominant development*, suggests an objective or linear perception in which time is constantly running out: With each moment, the past increases and the future decreases. Conversely, time expanding as one moves toward the future, so-called *future-dominant development*, suggests that personal growth and development affect one's perceptions of time. The past becomes less significant because one has already lived it; the future is the most significant time zone because it has not been experienced.[17]

For a configuration to represent temporal development, its circles must be of three different sizes, with the present circle the second largest. If the past circle is the largest, and the future circle is the smallest, the total configuration represents *past*-dominant development. In contrast, if the future circle is the largest and the past circle is smallest, the configuration represents *future*-dominant development. As we noted earlier, the configuration drawn by a majority of the corpsmen and corpswomen (Figure 7) was the future-dominant development. This configuration also happened to be atomistic but we do not yet know whether there is a relationship between the temporal development variable and the temporal relatedness variable.

The results shown in Table 26 indicate that for women, integrated and projected designs were associated with temporal development ($\chi^2 = 6.01$, $p < .05$). The table makes no distinction between past and future dominant development, but the correlations between development and future dominance ($r = .61$ for men; $r = .68$ for women) suggest that development is associated with future dominance: The circle representing the future is the largest, while the circle representing the past is the smallest.

Although for some people, the size of circles on the Circles Test may represent nothing more than a span of actual time—a number of years, for example—the fact that our respondents drew atomistic time zones and the future-dominant development configurations more than other configurations (Figure 7) confirms a finding by Florence Kluckhohn that American values emphasize the future and deemphasize the past. The configuration also supports Edward Hall who observed, "Not only do we Americans *segment* [my italics] and schedule time, but we look ahead and are oriented almost entirely toward the future."[18] Still, to connect cultural values to personal perceptions of the future can be misleading because human beings maintain perceptions of the past and present. That future dominance prevails on the Circles Test probably means little more than that the majority of this particular respondent population is "oriented almost entirely toward the future" and that cultural values have some affect, at least, on this group's perception of time. But we must underscore Hall's words, "almost entirely," for the Circles Test has not confirmed that cultural values determine personal perceptions of time. The sizes and interrelationships of the circles do not represent precisely, nor do they accurately express one's cultural values.

For the moment our important finding is that men and women who drew a continuous configuration on the Circles Test did not draw future-dominant development configurations as often as those men and women who drew either atomistic or integrated configurations. That past-, present-, and future-dominance configurations were unrelated for women suggests that women do not perceive the significance of one time zone in terms of the significance of the other time zones. Instead, women evaluate the zones one at a time. Furthermore, the absence of relationships among past, present, and future dominance supports the assertion that circle size represents something other than a mere span of time, a number of years. Still, to assert that circle size represents significance of a time zone dominance rather than a span of time raises the issue of the validity of the Circles Test. Does circle size represent dominance or duration?

THE VALIDITY OF THE CIRCLES TEST: DOMINANCE OR DURATION

Do circles drawn close together accurately depict a person's perception of linear continuity between time zones? Does the size of one circle relative to the size of another measure adequately how a person perceives the significance of time zones?

One measure of validity was to ask respondents to describe their reactions to the test. What perceptions of time did they want to express in their circles? Most frequently they said that proximity of circles was meant to convey relatedness and that circle size expressed their feelings about the significance of a time zone, not a chronological span of time.[19] A sensitive 12-year-old boy, who did not actually take part in the study, provided one interpretation of circle size when he said, "The past must be smaller than the future, for it is completed; it's over, while the future is forever."

As a second measure of the validity we compared relative circle size with responses to the Duration Inventory, in which we had asked people to define the chronological span of time zones (Chapter 4). We reasoned that if we found a high degree of association between chronological extension and relative circle size, we would have evidence that circle size represents chronological span more than the perceived significance of time zones.

The results of these correlations (Table 27) reveal two things. First, the size of the circle representing the past on the Circles Test was not significantly related to any of the Duration Inventory responses.[20] Second, for men the size of the future circle and the present circle were unrelated to any of the present and future duration variables. Among men a negative correlation was found between the duration of the distant past and present circle size ($r = -.21, p < .01$), and a positive correlation was found between the duration of the distant past and future circle size ($r = .28, p < .01$). Nonetheless, in general the results of the Circles Test do not seem to be highly related to the results of the Duration Inventory.

An unexpected correlation appears in Table 27. Temporal relatedness on the Circles Test was negatively correlated with the past duration of the present on the Duration Inventory for women ($r = -.21$) and with the near future duration of the present for men ($r = -.27$). Thus, the farther back in time women reach for experiences to incorporate into what they define as the present's duration, the more atomistically do they draw time zones on the Circles Test. This finding sheds a new light on the meaning of temporal atomicity. It suggests that by extending the past boundary of the present—the period of time into which recent events are incorporated- —women disengage themselves from the past. In contrast, as men extend the *near future*—the period of time including soon-to-occur events—the more they draw the circles atomistically.

Returning to the other significant correlations in Table 28, we ask why for men does an increase in the chronological extension of the distant past tend to decrease *present* dominance ($r = -.21, p < .01$) but *increase future* dominance ($r = .28, p < .01$)? One way to answer this is to look at the affect of the perception of historical time on the chronological duration of the present. Compared to the enormous stretch of time we call the historical past—the past prior to our birth—the present is insignificant. To the men in our sample, however, the future's significance (future dominance on the

Circles Test) is related to the chronological extension of, and presumably to the significance of, the historical past. Perhaps the conception of future dominance, a conception that Hall claims is held by most of us,[21] gains its credibility from one's acknowledgement of historical time.

In a letter written as a response to a study in which an investigator had related age to the perceived acceleration of time,[27] Benford maintained that time seems to pass faster as one gets older not because with each moment we are nearing death, but because the experiences accumulated in one's memory continuously increase.[22]

The implication of Benford's argument is that one assimilates time through two discrete modes: through memory and anticipation and through the immediate experiencing of events and feelings. Not only are these modes clearly related, they also involve linear and spatial conceptualizations of time. Yet one's preoccupation with memory or anticipation may reduce the intensity of one's preoccupation with immediate events and experiences. If this is so, the continuously increasing store of memory mentioned by Benford becomes associated with the ways in which we imagine the future, namely, anticipation and expectation.

SPATIAL VARIATIONS AND THE AFFECT OF MEASURED INTELLIGENCE

The straightforward scoring procedures used for the Circles Test do not do justice to the richness and imagination of our respondents' perceptions. We did not, for example, devise a way to account for a certain shading of the circles which gave a zone a three-dimensional quality. Furthermore, our scoring procedures for temporal relatedness, dominance, and development did not account for variations—for example, time zones intentionally drawn to fall off the edge of a page, or circles that had been flattened into the shape of eggs or hotdogs. More important, the scoring procedures ignored spatial references. We made no distinction, for example, among circles ordered from left to right, up to down, lower left to upper right, or circles drawn so that the present, say, sits on top of the past and future.[23]

That we did not consider these variations is, of course, our loss. Each of them represents a perception of time that would have been worth exploring. One young woman, whom we did interview in some detail, drew the past and present on one horizontal plane. She drew the future on a noticeably lower plane but still overlapped the future with the present. Asked to interpret her design, she said she hoped the future would bring recognizable change and somehow "steer her off her present course." By drawing overlapping circles, she retained her awareness of temporal evolution. More important, she communicated ingeniously that she associated external change with some personal stability—changelessness. Her design, therefore, represented the dynamics of what Erik Erikson has called identity. In large measure identity is the sense that certain aspects of one's self are changing, while a certain inner core of oneself remains unchanged.

The variations in the Circles Test evidenced by the young men and women in our sample raises a more serious issue, however, than the inadequacy of our scoring procedures. Specifically, we must ask whether intelligence or a special quality of thoughtfulness is associated with perceptions of temporal continuity or integration? To answer this question, we correlated intelligence scores[25] with the temporal relatedness scores, but no significant values were found ($r = .10$ for men; $r = .08$ for women).[26]

CONCLUSION: THE AFFECT OF COMPETENCE ON PERCEPTIONS OF TIME

The Circles Test differentiates people according to the time zone they deem most significant. Some people perceive the present time zone as being insignificant, merely a period of preparation for the future. This might symbolize what Kluckhohn called a *becoming* orientation,[27] an orientation in which one perceives the present as a bridge. In contrast, when the present is *most* significant it symbolizes Kluckhohn's *being* orientation. In this, people invest their energies in ongoing activities for the sake of the activities themselves—in and for the present, not the future.[27]

The Circles Test also differentiates people according to the relationships they perceive between the past, present, and future. For some persons, the past, present, and future are disconnected. They are united by nothing except sequence. This perception is called *atomistic*. For other persons, the zones are visualized linearly. Moments follow moments, and each moment, once experienced, disappears irretrievably into the past. Similarly, each experience has its own finite moment. This perception is called *continuous*. For still other people, individual moments seem to be extended. The past, present, and future appear to overlap—a perception called *integrated*—or occupy two zones simultaneously—a perception called *projected*. More than any others, these perceptions typify the meaning of the spatial conceptualization of time flow.

Why some people perceive time as integrated or projected may be partially explained by the theory of competence. For Robert White, the sense of competence develops from experiences that teach people that their personal *effort* or *activity* produces some desired *predictable effect*.[28] Through repeated experiences of this sort one begins to believe that if past efforts have shaped the present, then present efforts can shape the future. Although not referring directly to perceptions of time, White's words are relevant to our understanding of it. "If we conceive of structure as competence, we are giving it the dynamic character of patterns of *readiness* for future action" [emphasis added].[29]

With this notion of competence, we may speculate that the part of the past that overlaps the present symbolizes recalled experiences that shaped the present. Similarly, when the present and future overlap, we find symbolized a belief that the future can be influenced, caused, or controlled through present activity. Finally, from the configuration in which the present is totally *projected* into the future, we find the suggestion that present activity affects *all* that the future holds.[30]

In this competence model people make inferences about the future on the basis of their expectations of what the future will hold. Childlike hopefulness is not guiding their conceptions of the future; they are making plans, believing strongly that they can shape the future through their own efforts. In a sense expectations are like kites. They sail out beyond us, and although some people contend that in time we will reach the kites, others actually feel the string—that is, they feel connected to their expectations. This is the basis of White's notion of competence, a notion reminiscent of Edward Tiraykian's sentiment: ". . . and this leads me to realize that I am not contained in my finite and solid appearance but that my being goes out spatially and temporally."[31]

One may also wish to speculate that in the atomistic conceptualization of time the spaces separating the present from the past and future represent what Kurt Lewin called an *interference* in the emergence of competence.

If the free play activity of a child is interfered with, his average level of productivity may regress. . . . This regression is closely related to the child's time perspective. . . . He feels himself to be on insecure ground; he is aware of the possibility that the overwhelming power of the adult may interfere again at any moment. . . . This has a paralyzing effect on long-range planning.[32]

Following Lewin's argument, the person's perceptions of a relatedness between the past and the present may be interfered with. Unable to perceive a connection between them, one cannot recognize that certain past experiences caused the occurrence of certain present experiences.

The competence model makes one more contribution to our understanding of performance on the Circles Test. If circle configurations are meant to depict the expression of prior and anticipated action, the experience of change may influence personality factors such as anxiety, the need to achieve, cognitive styles, and conceptions of self.[33] Time, therefore, becomes for the behavioral scientist what it was for the philosopher Ernst

Cassirer: [That] all-embracing form for all changes: as a universal order in which every content of reality "is" and in which an unequivocal place is assigned to it . . . [for] all combinations of things, all relations prevailing among them, go back ultimately to determinations of the temporal process, to divisions of the earlier and later, the "now" and the "not now."[34]

With the Circles Test we encouraged our respondents to conceptualize the passage of time spatially. The Circles Test stressed the perception of the relatedness of time zones. Indeed, the configurations we called integrated and projected showed time zones residing within one another. In the atomistic conceptualization the time zones are not perceived as an unbroken line, but rather separated, disconnected.

The continuous configuration, however, is the one that comes closest to representing the linear conceptualization of time flow. Moments succeed moments, and no moment can reside within another moment. In the following chapter we turn our attention to an instrument that, unlike the Circles Test, encourages respondents to think about the passage of time in linear terms. With the use of this test, we examine further a conceptualization of continuous time flow.

NOTES

1. Friedrich Kummel, Time as Succession and the Problem of Duration. In *The Voices of Time*, J. T. Fraser (Ed.)(New York: Braziller, 1966), p. 40.
2. Heidegger. *Being and Time* (New York: Harper & Row, 1962).
3. Bergson referred to this form of relatedness between time zones as a concrescence of time zones.
4. On this point, see H. Zentner, The Social Time-Space Relationship: A Theoretical Formulation. *Sociological Inquiry*, **36** (1966): 61–79.

5. McClelland, *Personality* (New York: Holt, Rinehart and Winston, 1951), p. 486.

6. Cited in Rollo May, Contributions of Existential Psychotherapy. In *Existence*, Rollo May, Ernst Angel, and Henri Ellenberger (Eds.) (New York: Basic Books, 1958), p. 66.

7. Fraisse, *The Psychology of Time*. New York: Harper and Row, 1963, page 151.

8. On this point, see Timothy Leary, *Interpersonal Diagnosis of Personality: A Functional Theory and Methodology for Personality Evaluation* (New York: Ronald Press, 1957).

9. Much of the work on the Circles Test was influenced by the writings of Georges Gurvitch. Not only did we draw upon many of his concepts, we have also used several of his terms, although in doing this, we have perhaps violated his original definitions of these terms. See Gurvitch, Social Structure and the Multiplicity of Times. In *Sociological Theory, Values, and Sociocultural Change*, Edward Tiryakian (Ed.) (Glencoe, Ill.: Free Press, 1965), pp. 171–84, and The Spectrum of Social Time (Dordrecht, Holland: D. Reidel, 1964).

10. Heidegger, *op. cit.* The concept of temporal integration has also been discussed in another context. See Alden E. Wessman and David F. Ricks, *Mood and Personality* (New York: Holt, Rinehart and Winston, 1966).

    The variables generated in the Circles Test only barely resemble those processes described by Gurvitch. Consider, for example cyclical time, which Gurvitch attributed to "archaic societies where the mythological, religious and magical beliefs play such an important part." On the Circles Test, a configuration like that shown in Figure 3 might represent an expression of "cyclical time," but we call it temporal projection and therefore stress Gurvitch's description of "the past, present and future [as being] mutually projected into one another." See Gurvitch, *Spectrum of Social Time*, p. 32.

11. See, for example, Kluckhohn and Strodtbeck, (Eds.), *Variations in Value Orientations* (Evanston, Ill.: Row, Peterson, 1961).

12. Cited in Fred Weinstein and Gerald M. Platt, *The Wish To Be Free* (Berkeley: University of California Press, 1969).

13. This signature is borrowed from Kluckhohn and Strodtbeck, *op. cit.*

14. The actual frequency distribution of relatedness designs is shown in Table 22. A slight degree of variation in relatedness scores between the sexes is revealed in their standard deviation values (male, s.d. = 5.90); female, s.d. = 5.12). The difference however between the mean scores is not significant ($t = .41$).

15. In correlations of this sort variables are scaled so that the higher the number, the greater the relatedness or relative dominance of a particular zone. Women, therefore, either reduce the importance of the future as they integrate time zones or render it greater importance as they separate it off from the past and present.

16. In this computation dominance and relatedness scores were treated as categories rather than as scales. Zero represents atomicity, 2 represents continuity.

17. One might be tempted to think of future-dominant development as a life orientation and past-dominant development as a death orientation, but such attributions cannot be justified on the basis of the present data.

18. Edward T. Hall, *The Silent Language* (Garden City, N.Y.: Doubleday, 1959), p. 29.

19. This finding was not surprising because a previously administered instrument, the Lines Test, had explicitly instructed subjects to mark off purely chronological divisions of time. This test is described in Chapter 8.

20. The variables shown in Table 26 are a mixed bag of scales. Duration responses range from seconds to years; relatedness from a score of zero (atomicity) to the six points of

concentric circles (projection). Time zone size is measured relative to the other two zones. For the purpose of including the development variable in a correlation matrix, designs received a score of 2 if either past-or-future-dominant development was present and 1 if development was absent.

21. Hall, *op. cit.*

22. F. Benford, Apparent Time Acceleration with Age. *Science*, **99** (1944): 37.

23. Protocols from a sample of Indian students of identical age as the hospital corpsmen and corpswomen and comparable social class often revealed the past inserted *between* the present and future or the future between the past and present. These variations, however, never occurred in the American sample.

24. See Erik H. Erikson, Identity and the Life Cycle. *Psychological Issues*, **1** No. 1 (1959); see also his *Young Man Luther* (New York: Norton, 1958).

25. Intelligence scores were based on the Army General Classification Test administered to all entering military personnel. Intelligence represents the sum of verbal and mathematical aptitude scores which are themselves positively correlated (male, r = .51; female, r = .46).

26. A cross-tabulation between the various forms of relatedness and high, medium, and low intelligence groupings was also computed to learn whether the intelligence scores among those persons drawing continuous designs vary significantly from the intelligence scores of those persons drawing atomistic and integrated designs. Again, no significant levels of association emerged (male, χ^2 = 3.69; female, χ^2 = 2.86).

27. Kluckhohn, Dominant and Variant Value Orientations. In *Variations in Value Orientations*, Kluckhohn and Strodbeck (Eds.), op. cit.

28. R. W. White, Motivation Reconsidered: The Concept of Competence. *Psychological Review*, **66** (1959): pp. 297–333.

29. R. W. White, Ego and Reality in Psychoanalytic Theory: A Proposal Recording Independent Ego Energies. *Psychological Issues*, Monograph II, **3**, No. 3 (1963): 186.

30. Temporal integration and projection designs also represent psychological forms of what Gurvitch called "cyclical time." See Gurvitch, *Spectrum of Social Time;* see also L. W. Hearnshaw, Temporal Integration and Behavior. *Bulletin of the British Psychological Society*, **30** (1956): 1–20.

31. Edward Tiraykian, The Existential Self and the Person. In *The Self in Social Interaction*, C. Gordon and K. J. Gergen (Eds.) (New York: Wiley, 1968). Minkowski too, expressed a similar thought: "In this personal impetus there is an element of expansion; we go beyond the limits of our own ego and leave a personal imprint on the world . . . this accompanies a specific, positive feeling which we call contentment—that pleasure which accompanies every finished action or firm decision." See his *Le Temps Vécu, l'Evolution psychiatrique.* Paris: Centre d'Editions Psychiatriques, 1933.

32. Lewin, *Resolving Social Conflicts* (New York: Harper and Bros. 1948).

33. In this regard John Muller wrote, "An organism brings about some effective change in its surroundings or in its relationship to them." See Muller, *Self, Time and Activity.* Unpublished manuscript, Department of Social Relations, Harvard University 1967.

34. Ernst Cassirer, *Philosophy of Symbolic Forms*, Vol. III, *The Phenomenology of Knowledge* (New Haven: Yale University Press, 1967).

CHAPTER EIGHT

A LINEAR CONCEPTION OF TIME:

THE LINES TEST

In the linear (objective) conceptualization of time, time is perceived as composed of moments, one succeeding another in an unending continuum. We experience them one by one, unable to retrieve prior moments or to experience future moments until they reach the present. Even if we perceive time in spatial terms, we cannot deny the linearity of moments; we cannot actually slow or hasten the flow of time, nor can we extract a moment from one place in the time flow and insert it into another place.

In contrast to this conception of time flow is the spatial (subjective) conception. Here, we feel that through the act of remembering we can return the past to the present, and through expecting we can sense what the future will be like. Mircea Eliade echoed Kierkegaard and Neitzsche when he said that all time is revived in the eternity of the moment, a statement that must assume a spatial conception of time, not a linear one.[1]

To examine the spatial and linear conceptions properly, we must ask how these two views of time help us understand the nature of time as we experience it. We begin to answer this question by again reviewing the writings of Henri Bergson.

Bergson believed that we are conscious of at least two forms of duration—two ways, that is, in which moments come together. He called actual chronological duration—clock or physical time—*homogeneous duration*. The form of duration of which we are consciously aware he termed *heterogeneous duration*. Here we subjectively perceive of moments as permeating one another.[2]

From these two forms of duration, Bergson developed two notions of the conscious self. The first is based purely on homogeneous space and time in which one's *sense* of self is conditioned by concrete spatial and temporal calculations. The second, a *fundamental* self, is based on a person's ability to blend experienced moments into a whole. To illustrate what he meant, Bergson asked us to visualize him walking down a street for the first time, receiving two types of perceptual sensations as he walks. He experiences new sensations that can never be new again. He knows that in his future walks on this street he will receive the same sensations, perceive the same events. The fundamental self must be capable of integrating the novel experience of the first walk with successive walks and always be able to recall and appreciate first experiences.

If today's impressions were absolutely identical with those of yesterday what difference would there be between perceiving and recognizing, between learning and remembering? . . . What I ought to say is that every sensation is altered by repetition and that if it does not seem to me to change from day to day, it is because I perceive it through the object which is its cause, through the word which translates it.[3]

Throughout our study we have observed that perceptions of time involve both linear and spatial notions. We may think about the linear flow of time in terms of clock and calendar units (objectively), but we may also think of the same flow of time with emotion (subjectively), often with great intensity. We have made this point repeatedly.[4]

A second point that we have only hinted at involves the question of whether people can truly be objective about their perceptions of time. Can one really view time strictly in terms of a succession of chronological moments? We could ask how many seconds are in a minute, and our respondents would answer 60, thus leading us to believe we have made an objective inquiry into the meaning of time. But no inquiry into time perception, regardless of how objective it is intended to be, could prevent respondents from giving subjective responses. If one were to ask, for example, how long a year is, some people might answer, "Very long," or "Not long enough." They would give a subjective response to an objectively stated question and a subjective interpretation of an objective fact. This distinction between the objective reporting of temporal things and the subjective experiencing of them is illustrated in John Cohen's remark, "A month is about four times the 'duration' of a week, a month ago is not felt to be four times as remote as a week ago."[5]

For a long time psychologists attempted to control subjective responses to questions about time by focusing their research on the linear aspect of time. For example, they designed experiments in which a subject was told to watch for a light to go on and then to push a button when he thought that the light had been on for two minutes. Results of such "objective" inquiries showed that estimates of time passage varied because of the experimental conditions, but they also varied because the subject's age, sex, and social background affected his feelings, which affected even his most rudimentary estimates of the passage of time.[6]

Several investigators have tried to examine the nature of subjective responses to inquiries about the linear conception of time; they have asked their respondents to represent the duration of various temporal units "as they seem" to be. Cohen, Hensel, and Sylvester, for example, gave children a piece of paper with a straight line running across it. The children were told to think of the line as representing the time from "birth to now" and were instructed to mark off specific chronological units, such as days and weeks.[7] Because time is represented as a line in this experiment, the children are being encouraged to think of the passage of time in terms of the linear conceptualization. By making some objective calculations, the children should mark off the length of a day on the line as one seventh the length of a week. Sylvia Farnham-Diggory, in a study of time perceptions among psychotic, brain-damaged, and normal children, told respondents that the line on the page signified the time from now until the future. She asked the children to mark off intervals on the line representing two weeks, three hours, and 80 years.[8] In a study of the perceptions of time among dogmatic people, Bonier asked his respondents to mark off intervals on the line representing the lengths of stories told to them.[9]

In all of these studies the researchers used the same method—a time line—to allow people to report their subjective estimates of temporal duration. By translating duration into linear terms—time as a line and specific units of time as intervals on that line—the lengths of intervals marked off on the line could be examined to study how long people *felt* a particular duration to be. The time-line technique encouraged people to think of time in linear terms, but to give subjective estimates of time passage.

In this chapter we use the time line technique to explore the perceived duration of several specific units of time: the personal past (time from one's birth to the present), the personal future (time from the present until one's death), the historical past (time prior to one's birth), the historical future (time following one's death), and the life time (the personal past, present, and personal future as one unit). To measure these units, we designed the Lines Test, consisting of a piece of paper with a line running horizontally across it from one edge of the paper to the other. Respondents were told to think of the line as representing the passage of time. They were asked to make four marks on it to represent the moment of their birth, the moment of their death, and the boundaries of the present—that is, where they felt the past ends and the present begins and where they felt the present ends and the future begins.[10]

Because the instructions of the Lines Test encouraged the respondents to think of time passage in linear or chronological terms, we expected they would represent the personal past and future as an actual number of years. In the light of longevity statistics 20 year olds should draw their futures about three times as long as their pasts. But it was not so easy to predict the length of their life times, because this estimation requires the respondents to differentiate the time of *their* lives from *all* time. Also, some people might interpret the time line as representing their entire life time while others might decide that the duration of the historical past should be drawn longer than one's own life time. Some may even wish to mark off their own life times as a tiny fraction of the entire line, representing the number of years of a lifetime as a minute fraction of all time.[11] For these people, the moments of birth and death would be placed closely together on the time line.

There is nothing unusual about this last perception because in historical terms the duration of a single life occupies a very small segment of time. Phenomenologically, however, the sense of a life time is far more complex. Some people, for example, imagine time to flow along, pulling them with it, while others feel that they remain fixed and only time moves. Paul Tillich wrote that human experience is extratemporal, because life seems to reside outside of time. Upon death a person is *returned* to time. Heidegger also contended that man cannot exist in time because of his wholeness. What, he asked, would be the meaning of existence "no more" (the past) or "not yet" (the future)? There cannot be a moment-to-moment line over which the course of one's life runs; there can be only moment-by-moment progression. If Heidegger's conception is valid, birth and death cannot be considered temporal events, for neither can be incorporated directly into one's experience.

Philip Merlan,[12] whose writing has helped to clarify these notions, adds that Augustine, too, although deeply immersed in the scholastic temperament, seemed unable to consider birth a temporal event. This belief led him to write:

I am loath to count it /birth/ as part of my present life because, as to the darkness of oblivion, it is like the one which I spent in my mother's womb. . . . What do I still have to do with that of which I recall no trace.[13]

Another problem for the person responding to the Lines Test is deciding how long to draw the present on the line. One is really deciding between an objective (linear) conceptualization in which the present is defined as the instant of immediate experience, and subjective (spatial) conceptualization in which the present may be defined in terms of emotional experience. Asking people to mark off the boundaries of the present is really asking them to mark off how long the present *seems* to be, because ultimately the present is defined in both linear *and* spatial terms.

Although little is known about how people subjectively estimate the length of the present, the data reported in Chapter 4 showed that our respondents measured the present in chronological units ranging from seconds and minutes (the instantaneous present) to months and years (the extended present). The difference between the instantaneous and the extended conceptualizations of the present must be taken into account in our evaluation of how people perceive the present's length on the Lines Test. Perhaps one should avoid speaking of past, present, and future orientations when respondents disagree on when each of these three important time zones begins and ends.

The instructions to draw four marks on the line in the Lines Test produce six different line segments for study—the life time unit, the personal past, the present, the personal future, the historical past, and the historical future.[14] These segments may be seen in Figure 8. Because the length of the test line is finite, the longer one draws his or her lifetime, the shorter

1   Birth
2   Past—present boundary
3   Present—future boundary
4   Death

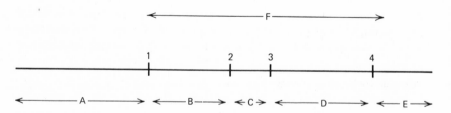

A   Historical past
B   Personal past
C   Present
D   Personal future
E   Historical future
F   Life time

Figure 8   Segments of the Lines Test.

the segment representing historical time will be, and vice versa. More important, *where* the person locates his or her lifetime on the line determines the length of the segments representing both the historical past and historical future. Respondents must decide how to scale their lifetimes, where to position their life times on the line, and then where to position the present.

How respondents will react to the Lines Test instructions is difficult to predict. Some persons may draw their personal past and personal future first and use whatever is left over on the line to represent the historical past and future. Others may first mark off the historical past and historical future and use whatever is left over to represent their life times. The life time, important as it is, may have to be shortened to acknowledge the enormous stretch of years constituting historical time.[15]

THE REALITY OF ESTIMATIONS

In her testing Sylvia Farnham-Diggory anticipated that her normal subjects would draw 80 years considerably longer than two weeks because the obvious disparity in their lengths cannot help but influence estimates of duration. Objectively, no one can argue with this. Prediction is more difficult when we ask people to draw on a line segments representing their lifetime or their personal past and personal future. For example, one's age should affect the perception of the length of one's life time. Young children indicate little interest in, or awareness of, historical time; so they would mark off a longer life time than would older people, who have knowledge of history and human evolution and would draw a shorter life time and a correspondingly longer historical past and historical future. But a man may use about half the line to designate his lifetime and leave the rest of the line to represent the historical past and historical future. While he has acknowledged the existence of the historical past and future, we must wonder whether he feels that the scale of his lifetime is a bit out of proportion. In working on the Lines Test he must make a distinction between what he sees as an appropriate linear representation of all time and how he subjectively compares with it the duration of his lifetime. This presents the conflict of time continuously running out on all of us, in linear terms, and time growing larger and more important as one grows older, in psychological terms. This feeling is reflected in Kurt Lewin's statement, "Growing through adolescence to young manhood or womanhood means enlarging the scope and the time perspective of one's psychological world."[16] In this context, then, the respondent's life time may seem far more important to him than historical time, even though he acknowledges that historical time is longer than his or anyone's life time.[17]

In evaluating the Lines Test the perception of the lifetime as being significantly longer than historical time (a nonspatial conception of time) is called an *egocentric* perception. In contrast, the perception of historical time as being significantly longer than the lifetime (a linear conception of time) is called the *historiocentric* perception.

SEX, SOCIAL CLASS, AND LINEAR EXTENSION

While many investigators have described the changes in how we perceive time as we develop from childhood to adulthood,[18] fewer investigators have researched how we perceive time as men or as women. Indeed, much research either omits one sex or the other or avoids discussing the possible relationship between gender and perceptions of time. In a review of the relationship between gender and the estimation of time passage, Paul Fraisse was convinced that one's sex somehow affected one's subjective estimations of time passage. He was puzzled, however, by these results and so was reluctant to interpret what he had discovered:

A critical study of the methods used does not explain these divergencies. We shall therefore not attempt to explain the possible differences between the time judgment of men and women. More experiments will have to be made in varied conditions to settle the question before it can be interpreted. We discussed it here only because we thought this preferable to passing over it in silence.[19]

Many psychological and sociological studies have demonstrated systematic and recurring differences between men's and women's attitudes, cognitive capacities, and actual behavior.[20] Some interpret these differences as distinguishing men's psychological capacities from women's psychological capacities,[21] while others say such variations are associated with social role patterning rather than psychological functioning. (In social role patterning men and women are taught to respond in terms of their sex, thus their attitudes and perceptions develop differently. For this reason, we use the terms *sex role* or *social role differences* rather than *sex differences.)*

Because of the many sex role differences, we must consider possible relationships between the characteristic social role patternings of men and women and performance on the Lines Test. For example, an interest in science is believed to be a characteristically male trait, so we might expect that men will prefer to represent the present in chronological terms, drawing the present on the Lines Test shorter than women will draw the present. Such a finding, however, would add little to our understanding of people's perceptions of the duration of time zones. But if we could turn up a series of findings distinguishing men's perceptions from women's, we might feel more confident that a person's sex role influences his or her perceptions of the length of the life time and the historical past and historical future.

Although the Lines Test seems simple on the surface, performance on it can lead to rather complex findings. For example, marking off a long personal future may indicate a future lasting many years, or it may indicate the future as a significant time zone. In the latter sense a long future on the Lines Test could be equated with future dominance on the Circles Test.[22] The combination of a long future and a short or instantaneous present may represent a life attitude devaluing the present because it serves as preparation for the longer and more important future. This combination could be compared with people rated future-oriented on the Experiential Inventory because their special concern with the outcome of events might lead them to draw short presents, signifying its insignificance, and long futures, symbolizing that their goals will be reached in the future.[23]

Although this discussion is speculation, it does demonstrate that a short present on the Lines Test may be a chronological estimate of the present, a perception that the present is insignificant, or other perceptions as well.

Similarly, interpretations of the long lifetime—the egocentric perception—are also difficult. The term *egocentric* does not mean self-centeredness or narcissism; it merely indicates that persons who mark off a long life time on the Lines Test are more involved with one or more time zones of their own life than they are with the historical past and future. We might expect that persons revealing an egocentric perception will be more involved with the present and future—the time zones of being and becoming[24] than with the past and will mark off larger time zones for the present and future and relatively short zones for the remaining time periods. Past accomplishments are thus quickly dismissed and present activities and future possibilities are magnified. The egocentric perception and its emphasis on directing one's present efforts on shaping the personal future and controlling personal outcomes in the future is somewhat reminiscent of what we called the orientation to achievement in Chapter 6. Achievement-oriented persons believe they can achieve goals on the basis of their own efforts.[25]

The historiocentric perception on the Lines Test represents something more than the chronological fact that one's life time is quite short relative to all time. It is also similar to an orientation described in Chapter 6—ascription,[26] an orientation toward what one inherits, both biologically and sociologically, as opposed to what one achieves through his or her own efforts.

The following example is an illustration of the possible relationship of the historiocentric perception to the ascriptive orientation. A rich man reports that the few years of his life can hardly be compared chronologically with the endless amounts of historical time. Yet this man's basic orientation to life may well be determined by the fact that past origins, traditions, and accomplishments of his family have provided him with his present position and security. Because he does not have to work hard for economic rewards, he may view the successful linking of past and future generations as the major goal of his life. He is oriented to the succession of generations rather than to personal accomplishment.

In terms of social class and time orientation, upper-class persons' orientation is toward *preserving* the family tradition, or *historical preparation*, which is in direct contrast with achieving or upwardly mobile persons who are oriented to *personal preparation* and change. Those in the latter group truly are the people belonging to the society described by Thistlethwaite: "
. . . a society fluid and experimental, uncommitted to rigid values, cherishing freedom of will and choice, and bestowing all the promise of the future on those with the manhood to reject the past."[27]

Several studies of social class and time perceptions conclude that the upper classes of Western societies are oriented toward tradition or the historical past, while the middle classes are oriented toward accomplishment, or the personal future.[28] If the upper-class view of life emphasizes the significance of keeping things as they are, the middle-class view of life emphasizes change and improvement. Important as these studies may be, they often fail to explore the differences in the time perceptions of men and women of the same social class. If we classify someone as middle-class or upper-class, we begin to suggest what sorts of time perceptions that person will have; if we know the person's age and sex as well, we have an even better starting point for predicting this person's perceptions of time.

Summary. External sources—age, sex, social class—influence perceptions of time: Perceptions change as one gets older; different perceptions are found in the middle and upper classes; and perceptions of men and women differ because of sex role. The term *sex role* underscores the point that during the process of socialization, people learn values, styles of interaction, and ways of understanding the world that eventually emerge as characteristically masculine or feminine.[29] Thus, we can begin to identify "typically" masculine and feminine perceptions of time.

We can predict performance on the Lines Test and the effect of external sources on the results by constructing a series of hypotheses:

1. Older persons will maintain a more realistic perception of time and a greater concern with history than younger persons; we called this perception historiocentric. Thus, on the Lines Test older people will draw shorter life times and shorter personal pasts and futures than younger people.

2. Upper-class men and women will be oriented toward what they inherit; we called this orientation ascription and related it to the historiocentric perception on the Lines Test. We expect these people to draw short life times on the test. Middle-class men and women will be oriented toward achievement, which we related to the egocentric perception on the Lines Test; their orientation will influence them to draw longer life times than those drawn by upper-class men and women. However, women, irrespective of their social-class origins, will be more concerned with generational continuity than men, and so they will draw shorter life times on the Lines Test.

3. Because they are oriented toward achievement, middle-class men will draw the future longer, but the present shorter, than upper-class men and women.

4. Because they are oriented toward achievement, middle-class men and women will draw smaller presents than upper-class men and women.

The Lines Test encourages respondents to think of the passage of time in objective (linear) terms. It represents time as a continuous flow, and respondents are asked to designate various periods of chronological time, for example, life time and the present. In contrast, our four hypotheses are based on the assumption that although people are instructed to think of time in objective (linear) terms, they give subjective responses, and these responses will be influenced by one's age, sex, and social class. Furthermore, a connection between these subjective responses and social structure is assumed.[30] If the data confirm these hypotheses, we can make a strong case that the Lines Test evokes subjective responses about the meaning of time zones.

RESPONDENTS AND PROCEDURE

Although much of the material presented thus far has hinted at the fact that people of various social classes perceive time differently, we have not yet empirically examined the effect of social class on perceptions of time. Indeed, as we noted in chapter 1, our test sample of corpsmen and corpswomen was carefully selected in order to obtain some degree of homogeneity in their social backgrounds so that we could look closely at the differences in time perceptions based on sex. To explore the influence of age, sex and social class on performance on the Lines Test, we will examine a different sample of people. The first set of results we will study are from the Lines Test administered to 180 students from two coeducational schools in Vienna, Austria. (Data from the American sample are discussed later in this chapter and in Chapter 9.)

The Austrian students ranged in age from 12 to 18 and were classified as upper or middle class on the basis of their fathers' occupation and the social position of their schools.[31] A breakdown of the Austrian sample is shown in Table 28. To compare the time perceptions of children as young as 12 with adolescents as old as 18 we divided the students into two groups on the basis of their age. We called students aged 12 to 15 young; students aged 16 to 18 were called old.

The Lines Test given to the Austrian students was a piece of paper with a 20-centimeters-long horizontal line drawn across it from edge to edge. The instructions were identical to those given to the American hospital corpsmen and corpswomen.[32]

RESULTS: THE AUSTRIAN CASE

The Six Variables. The means, medians, and range of all Lines Test responses are presented in Table 29; each number signifies lengths in centimeters. The Austrian students drew the personal future about three times longer than the personal past, indicating that they may have been using the Lines Test to signify chronological length of time zones. On the average they used 75% of the line to represent their life times, which would not be likely if their intention was to represent time zones in terms of number of years. Evidently, the subjective responses we expected on the Lines Test have emerged.

The young Austrians also drew short present extensions. They also drew historical time to mirror personal time: The historical future was longer than the historical past. No respondents extended the present outside of the life time, and all students used their birth as the origin of the past and their death as the conclusion of the future.

Correlations between the six variables of the Lines Test shown in Table 30 reveal the relationship between the length of the life time and the lengths of the historical past and future. Because of the nature of the Lines Test, the longer a person draws the life time, the shorter the space remaining for the historical past and historical future, and vice versa. Table 30 shows no relationship between the lengths drawn for the present and the life time (male, $r = .12$; female, $r = .19$) and no significant correlation between the lengths drawn by women for the present and the personal past ($r = .11$). However, there is a significant correlation between these two variables among men ($r = .24, p < .05$) and a negative correlation between the lengths of the present and the personal future drawn by men ($r = -.41, p < .01$).

We have discussed the complexities of perceiving the present's extension. The present may be drawn very short to represent an objective perception of it as the immediate instant of sensation, or it may be drawn very long to represent a more subjective perception, or how one feels about the present. The important point, however, is that there are no chronological guides to help the respondent determine the present's length. The personal past may be represented as the number of years a person has lived, and the personal future may be represented as the number of years a person expects to live, but the present may be perceived in terms of units ranging from microseconds to years. In the Austrian case the young men related the present's length with the length of both the personal and historical futures.

Life Time.   An analysis of variance was used to examine the relationship between age, sex, social class and the life time, as well as the other five variables on the Lines Test.[33] The analysis of variance for the life time is summarized in Table 31. The significant overall effect of social class on the life time ($F = 4.63$, $p < .05$) confirms the earlier hypothesis that upper-class students ($\overline{X} = 14.02$) tend to reveal the historiocentric perception; middle-class students ($\overline{X} = 15.43$) tend to draw the egocentric perception of the life time. The significant sex-by-class interaction ($F = 6.53$, $p < .05$) supports the hypothesis that middle-class girls ($\overline{X} = 14.24$) tend to be more historiocentric than middle-class boys ($\overline{X} = 16.48$). However, this difference is not due to any universal feminine historiocentrism, because upper-class boys ($\overline{X} = 13.39$) were more historiocentric than upper-class girls ($\overline{X} = 14.95$). Temporal egocentrism, therefore, occurs mainly among middle-class boys and upper-class girls.

The statistically significant sex-by-age interaction ($F = 6.34$, $p < .05$) (Table 31) shows that whereas older boys ($\overline{X} = 15.69$) tended to be *more* egocentric than younger boys ($\overline{X} = 14.02$), older girls ($\overline{X} = 13.62$) were *less* egocentric than younger girls ($\overline{X} = 15.52$). Thus, our earlier hypothesis that older persons would tend to draw shorter life times than younger persons is confirmed only among girls.[34]

Because the longer a person draws the life time on the Lines Test, the shorter he or she will draw the historical past and future, the life time is highly negatively correlated with both the historical past (male, $r = -.60$, $p < .01$; female, $r = -.84$, $p < .01$) and the historical future (male $r = -.86$, $p < .01$; *female, $r = -.92$, $p < .01$).[34] The analysis of variance for the historical past and historical future should demonstrate the same effects of age, sex, and social class as it did for the life time. Accordingly, no discussion of the results of the analysis of variance for the historical past and historical future is required.

The Present. In the analysis of variance for the present extension shown in Table 32, only the effect of age reaches statistical significance. We did not hypothesize the effect of age on the present's extension, but the test results show that older students ($\overline{X} = 1.24$) drew shorter presents—the perception we called the instantaneous present—more often than younger students did ($\overline{X} = 1.90$; $F = 4.30$, $p < .05$). We did hypothesize that men ($\overline{X} = 1.29$) would draw shorter present extensions than women ($\overline{X} = 1.84$), but as we see in Table 32, this hypothesis is only minimally supported by the data ($F = 3.27$), $p < .10$). The hypothesis that middle-class men and women would draw smaller present extensions than upper-class men and women is not confirmed by the data ($F = 1.49$).

Our discovery of a relationship between age and the present's extension—the only significant relationship between the length of the present and age, sex, and social class—is especially noteworthy in light of the earlier finding that the length of the life time drawn is not affected by age (Table 31). The older Austrian students drew the personal past and future longer than the younger students did to compensate for drawing a shorter present than the younger students did. This finding suggests a revision in Lewin's assertion that development from adolescence to young adulthood means "enlarging the scope and time perspective of one's psychological world."[36] Rather than enlarging the time perspective of early adolescence, development involves *reorganizing* it, for example, reducing the length of the present.

The fact that age, sex, and social class have different effects on perceiving the length of the present and the length of the life time is not surprising in view of the observation that the length of the present drawn is not related to the length of the life time (Table 30).

The Personal Past and Personal Future. A summary of the analyses of variance of performance on the personal past extension of the Lines Test is shown in Table 33. The personal past was drawn significantly longer by older students $(\overline{X}= 4.00)$ than by younger students $(\overline{X}= 3.41; F = 5.07, p <$ .05), reversing our hypothesis that older students will draw shorter personal pasts than younger students. In addition, the sex-by-age interaction effect that we observed in connection with the life time is reflected in the personal past $(F = 4.95, p\ .05)$ but not in the extension of the personal future. These results indicate that there is a significant sex difference in the alleged reorganization (a shortening of the present and a lengthening of the personal past) of the temporal perspective occurring between early and late adolescence. As boys get older they draw *longer* personal pasts on the Lines Test (young boys, $\overline{X} = 2.91$; older boys, $\overline{X} = 4.12$); as girls get older they draw slightly *shorter* personal pasts on the Lines Test (young girls, $\overline{X} = 3.97$; older girls, $\overline{X} = 3.88$).[37]

The results of the analysis of variance of the personal future shown in Table 34 indicate that middle-class boys $(\overline{X}= 11.2)$ drew longer personal futures than upper-class boys $(\overline{X}= 8.47)$, middle-class girls $(\overline{X}= 8.12)$, and upper-class girls $(\overline{X} = 9.63)$. This finding is significant statistically $(F = 10.91, p < .01)$ and supports our third hypothesis—middle-class boys, because they are oriented to achievement, will draw a longer future than will upper-class boys and girls.

The Historical Past and Historical Future. The longer one draws the life time, the shorter the line segments for the historical past and historical future. If middle-class boys and upper class girls draw egocentric designs on the Lines Test, middle-class girls and upper-class boys draw historiocentric designs. Is the egocentricity among middle-class boys and upper-class girls due to the fact that they have drawn a shorter historical past, a shorter historical future, or have shortened both of these time zones?

The analysis of variance of the historical past and historical future indicates that middle-class boys, in particular *older* middle-class boys, drew shorter historical futures than upper-class boys. Furthermore, older boys drew shorter historical future extensions than younger boys ($p >$ .001). These results indicate that middle-class boys are egocentric because they increase their life time by shortening the historical future, but not by shortening the historical past.

Combining these results with our earlier results shows that middle-class boys increase their life times by shortening the historical future and lengthening the personal future. In contrast, the historiocentric middle-class girls draw their life times short and both the historical past and historical future long.

LIMITATIONS AND IMPLICATIONS OF THE LINES TEST

The major purposes of the Lines Test were to demonstrate that subjective responses are as important in perceiving time zone extensions as are objective estimates and that these subjective responses, which at first glance may seem highly personal and even idiosyncratic, are influenced by one's age, sex, and social-class background.[38] The data collected on the Austrian school children support these two points. Indeed, only a few of the findings shown in Tables 31 through 34 (for example, the performance of upper-class boys on the personal past extension) could be considered objective (chronological) rather than subjective estimates of a number of years.[39]

In interpreting the findings on middle-class time perceptions we should recall that men and women of the middle class are characteristically oriented to achievement and social mobility.[40] Theoretically, middle-class persons believe that success in life implies that one be something better in the future than one is at present. Such beliefs are fundamental to the concept of social mobility and to what more recently has been called social upgrading. Some middle-class men and women prefer to think of their development as a continuous process in which one is barely able to see himself or herself changing; other middle-class men and women think of their development as a discontinuous process in which they perceive the present as wholly distinct from the past. Eventually this second group feels themselves to be totally different from what they once were.[41] In either case middle-class men and women will perceive the future as the most important time zone.

Given these orientations, we expect that middle-class men and women would draw long life times on the Lines Test—the egocentric perception—and long personal futures. We expected moreover, to find middle-class boys drawing longer personal futures than middle-class girls because boys are more socialized than girls toward creating a better status for themselves in the future and they are also taught that the personal past and future are more important than the historical past and future. The personal future contains life's significant goals, and the personal past is the base line against which they measure the extent of their accomplishments and progress. Middle-class boys must relinquish their pasts and deemphasize past achievements.[42] This point is reflected in Thistlethwaite's words, " . . . bestowing all the promise of the future on those with the *manhood* [emphasis added] to reject the past."[43]

Upper-class men and women are oriented to generational descent and family succession. An important emphasis is placed on inheritance of legacies and material goods; thus, preservation of what one already possesses is as important if not more important than personal advancement and change. The perception of a historiocentric life time represents not only an appreciation of historical time, but in addition, an appreciation of family *tradition* and the passing on of a family's name and its worldly goods. Personal accomplishment and achievement are important, but the upper class evaluates personal achievement both in terms of how well an individual has done in the world and how well the individual has been able to maintain the traditions of his or her family and class.[44]

While middle-class career aspirations depend on individual advancement, upper-class aspirations focus on the protection of family name, family history and family members. The perception of an historiocentric life time, therefore, suggests not merely the awareness of past history and the contemplation of a destiny based on the past, it suggests an acknowledgement of how wealth and position are inherited from one's ancestors, or more generally, how wealth and position contribute to upper-class personal achievement and to one's position in the society.[44]

The results of the Lines Test for the Austrian school children show that although the middle-class boys, as we expected, drew significantly long personal futures, middle-class girls drew long presents. So there is a sex role difference in the middle-class perception of the present and future: Middle-class boys perceive the future as the most important time zone; middle-class girls perceive the present as the most important time zone. Perhaps the extended present on the Lines Test signifies that middle-class girls believe in the linear perspective of time: Because one lives only in the present, it is seen as the most important time zone. The extended present on the Lines Test may also indicate a stabilizing role for the woman in western society.[47] This demand for stability and constancy in a woman may create the feeling that not only does she live in the present, as the linear perspective emphasizes, but also that her future will be little more than a continuation of the present, if not a duplication of it.

There is a clear meaning in the middle-class girl's Lines Test: She refuses to define the present as simply the flow of mechanical, clock-ticking seconds. Her historiocentric life time also indicates her awareness of the contribution she as a mother must make to the succession of generations. But these orientations to tradition and the succession of generations are not identical to those held by upper-class boys and girls. More likely, middle-class women bear children not to perpetuate family name and tradition but because society expects them to become mothers and because personally many want to become mothers.[48]

Social class and sex role affects perceptions of time; it also affects performance on the Lines Test. If the upper class perceives time in terms of the importance of the next generation maintaining the traditions of prior generations, perhaps the upper class prefers to think of time in linear terms. After all, in linear terms, one imagines a line of time in which moments succeed moments. In some of these moments a person is born, in some of these moments a person dies. In effect, this is the basis of both linear succession and family succession. Someone dies and someone else carries on in his or her place. Time in this linear perspective becomes not only an ineluctable flow, but an ineluctable flow of lifetimes.[49] Moreover, the instructions of the Lines Test suggest this flow of life times- —respondents are asked to mark off the moments of their births and deaths.

Upper-class traditionalism may encourage people to think about time in terms of the succession of generations, but it is not as clear that middle-class socialization demands that people think of time in linear terms. Middle-class boys, who theoretically are advocates of progress and change, cannot allow themselves to believe that events merely follow one another in time. They must convince themselves that events occur because one *causes* them to occur. The hero in John Schultz's novella *Custom* makes this very discovery about time. "I became certain that time is properly measured only by the size of the effort, the amount of spent energy." Perhaps the Lines Test encourages middle-class men and women to think of time in terms that are not particularly congenial to them. The Circles Test, which encourages people to think of time in spatial and subjective terms, might be a more congenial instrument to members of the middle class.

DEVELOPMENTAL PERCEPTIONS OF TIME

Age, sex, and social class affect a person's perception of time, and these variables occasionally interact with each other. If we speak of the effect of one variable on one's linear time perceptions, we must know the other two variables.[51] We cannot argue that older middle-class boys who draw a long life time on the Lines Test, are less aware of historical time than older middle-class girls or upper-class boys and girls. A better argument would be that the overriding achievement strivings of older middle-class boys influence them to think of their own life times as the most important segment on the Lines Test. Thus, middle-class boys draw long life times because their life times are important to them and not because they are unaware of the existence of the historical past and future.

Further research on the topic of age variations in perceptions of time may confirm the notion that children learn about the meaning of time during two distinct periods of socialization. During the first socialization period, roughly during the years 5 through 8, children learn that other people have personal histories and have experienced these histories long before the time of the children's own birth. The child develops a sense of how each person in the world lives his or her own life differently from the way the child leads his or her own life. This period is characterized by the transformation of egocentrism to allocentrism.[52]

The second socialization period occurs around puberty, when careers or future commitments generally require young adolescents to alter their perceptions of time according to the values, traditions, and customs of the society. At this point adolescents' conceptions of the future change because they are learning what realistic possibilities exist in the future. In addition young adolescents must learn the difference between what *they* might want to happen in the future and what their parents and teachers might want to happen in the future.[53]

Presumably, no manifest changes in time perceptions may occur during this second stage as was the case with middle-class girls' perceptions of the present. Yet perceptions at this time may also revert back to the egocentric forms that existed prior to the first stage, as was the case with middle-class boys. Future research will have to distinguish between the forms of early childhood egocentricity, which represents highly solipsistic perceptions, and the later forms of egocentricity, which symbolize legitimate achievement values.

One might also wish to argue that such perceptual shifts are a result of varying intelligence levels. Because intelligence scores were not available for the Austrian students, nothing definite can be said about this matter. Yet even if intelligence levels were higher among upper-class students, nothing more is explained, because middle-class boys would then be responding "intelligently" in terms of their present extensions but not life time extensions, and middle-class girls in terms of their life time but not present extensions.

CHRONOLOGICAL CONFIRMATIONS AND THE TIME OF IMAGINING: THE AMERICAN CASE

To confirm the notion that many of the Lines Test variables were representations of a period of years, the American corpsmen and corpswomen were instructed to write a brief account of their reactions to the Lines Test. Over 98% of the men and women reported that they were representing years on the test in drawing the life time and also in drawing past, present, and future extensions.[54] Without question, the Lines Test had encouraged our respondents to think of the passage of time in chronological (linear) terms.

As in the Austrian sample, American mean scores revealed that the extensions of the personal past and personal future occupied almost 3/4 of the line. Similarly, the Americans drew the historical past and historial future almost identical in length to those drawn by the Austrians. About

7% of the Americans drew their own life time two centimeters long or less to show their belief that a single life occupies a very small amount of time relative to all time. About 20% of the American hospital corpsmen and corpswomen used the entire line to represent their life time, revealing a perfect egocentric perception of the life time. On most points, however, the American performance on the Lines Test was essentially the same as that of the Austrian respondents.

Although there was considerable variation in the distribution of American scores, one point of similarity was the fact that many persons drew the historical past and historical future extensions the same way they had drawn the extensions of the personal past and personal future. They drew the historical future longer than the historical past and the personal future longer than the personal past. There are at least three possible reasons for this result. First, individuals may have imposed their definitions of personal time onto the total extension of time. Second, some persons may have wished to communicate the belief that while the world has existed for an enormous amount of time, there was in fact a starting point, a point where time began. Presumably, the period following death, a period we often refer to as forever, seems longer than the historical past is now or will be in the future.[55] Third, longevity statistics provide a mathematical baseline for determining one's life time, but not for determining the length of historical time. Thus, the length of historical time one draws is associated with a person's acknowledgement of history.

Variations in extension of the historical past and historical future suggest a possible relationship between an interest in recovering the historical past and knowing the historical future and how long these periods are drawn on the Lines Test. Are there positive correlations between the extensions of historical time on the Lines Test and the amount of money offered in the Money Game for the historical past and future?

The correlations shown in Table 35 involving American students suggest a positive association between the extension of the historical past on the Lines Test and a desire to recover time from the historical past on the Money Game. However, a comparable relationship does not exist for the corresponding historical future variables. In fact, although not attaining statistically significant levels, four of the six correlations in Table 35 reveal negative values. Thus the only positive correlation involving the historical past is that the longer it is drawn on the Lines Test, the greater is the interest in recovering it on the Money Game, and vice versa.

LIFE TIME AND THE TYPOLOGIES OF EGOCENTRIC AND
HISTORIOCENTRIC TIME

We tried to confirm the notion that the sizes of the time zones on the Circles Test were drawn to represent the perceived importance of the time zones rather than the amount of chronological time defining the time zones. To do this we correlated circle size on the Circles Test with the measurements of the perceived extension of the past, present, and future on the Duration Inventory. We found some significant correlations that tended to confirm the notion that circle size was not simply a representation of chronology.

With the results of the Lines Test analyzed, we can also confirm the notion that circle size represents the perceived significance of time zones by correlating results of the Circles Test and Lines Test. If circle size does not represent chronology, as the extensions on the Lines Test seem to, we should find no statistically significant correlations among the variables of these two tests. In fact, the correlations shown in Table 36 indicate that with two exceptions, the Lines Test and Circles Test results are almost wholly independent. The two exceptions are important because they add to our understanding of the meaning of present dominance. Table 35 shows that only among women is present dominance related to a shortened life time ($r = -.20$, $p < .05$) and a longer historical past ($r = .27$, $p < .01$).

In addition to these correlations we may investigate the relationships between the Circles Test and Lines Test in still another way. We can divide the American respondents into three groups: those who drew especially egocentric life times, those who drew especially historiocentric life times, and those who drew average life times (the middle group).[56]

Very few of the American students drew a perfectly historiocentric life time. The values in Table 37 show that the historiocentric perception includes drawings that devote almost one-half of the total line to history. Yet even with this large allotment of the line given to historical time, only about 25% of the hospital corpsmen and corpswomen are represented. Evidently, we are dealing with a skewed distribution in which most people drew long life times, the perception we have called egocentric.

If we now compare performance on the Circles Test by these three groups, we find that temporal development on the Circles Test is associated with drawing an egocentric life time on the Lines Test in men ($\chi^2 = 4.61$, 2 df) and women ($\chi^2 = 4.18$, 2 df). The finding for men is especially interesting because we found in Chapter 7 that temporal development is characteristically a feminine perception.[57]

This finding raises the following questions: Does historiocentrism on the Lines Test reduce the size of the present circle on the Circles Test, or does it make the present so insignificant that it is drawn as the largest circle (present dominance)? Does historiocentrism disrupt temporal development on the Circles Test and encourage future oriented perceptions of time generally?[58]

The percentage distributions in Table 38 show that men and women who drew long (egocentric) life times on the Lines Test were also likely to draw the present as the second largest circle on the Circles Test. This perception is called secondary dominance. Those who drew short (historiocentric) life times were also likely to draw the present as the largest circle on the Circles Test. This perception is called present dominance.[59]

In our discussion of the Circles Test we interpreted a present drawn larger than one time zone, but smaller than the other time zone (Figure 7 in Chapter 7) to mean that respondents felt the present to be important but not *the* important time zone. In fact, when the present is drawn as the second largest time zone, the future usually is drawn as the largest and most significant time zone and the past is drawn as the smallest and least significant time zone. Thus, we assumed in the future-dominant design that the present was important essentially as a period of preparation for the future. That the future is perceived as the most important time zone is not surprising, given the ages of the American and Austrian students. Yet, age alone does not adequately explain variations in perceptions of time.[60]

CONCLUSION

Dilthey once remarked that the future becomes the past so quickly that we cannot ever know the feeling of the present. To a degree, this observation explains the egocentric perception, which is characterized by a concern with personal growth and achievement. A sense of *becoming* in the present makes it a period of preparing for the future. Historiocentrism is characterized by a concern with historical time, concern with what one is experiencing in the present—the experiencing of *being*.[61] The present, therefore, is much more than just a time to prepare; in it one can experience the moment when a particular action is completed and can also experience the resulting satisfaction. For this historiocentric perceiver, the present presumably contains ingredients of the past and future in the form of memories and expectations, as Jerome Frank alleged when he wrote that the past, present, and future seem as one.[62] But this spatial conception of time does not imply that persons who draw short life times perceive the time zones to be more integrated than those persons drawing long life times. Indeed, the data from the Lines Test and Circles Test suggest that both egocentric and historiocentric persons think about the present as a combination of memory, anticipation, and involvement with ongoing events, experiences, and feelings.

Perhaps the future-oriented quality of the egocentric perception "pulls" persons who draw long life times forward in such a way that they feel disengaged or alienated from the flow of time. In contrast, historiocentrism, which is similar to what David Riesman called a "tradition directedness,"[63] predisposes one to perceive the present's fullness. People begin to recognize how their own development is similar to the history of cultures and societies, as well as to the histories of individual men and women. They then become more engaged in the experience of the moment and less concerned with present preparations for the future and future achievement.

This view can be called the optimistic side of the historiocentric perception; but a passage from Blaise Pascal reminds us of the frightening side of contemplating human finitude and the small amount of time each of us is alive:

When I consider the brief span of my life, swallowed up in the eternity before and behind it, the small space that I fill, or even see, engulfed in the infinite immensity of spaces which I know not, and which know not me, I am afraid and wonder to see myself here rather than there; for there is no reason why I should be here rather than there, now rather than. . . .[64]

The linear conception of time characteristically involves chronological estimates of the lengths of time zones. To study linear time perceptions of two test groups—Austrian students and American hospital corpsmen and corpswomen—we used the Lines Test. On a line representing the passage of time, respondents were asked to mark off a segment to show the period between their births and deaths (the life time line) and the boundaries of the end of the past and the beginning of the future (the present line). Despite wide variations in results, the unifying pattern was that the lengths of line segments reflected the age, sex, and social-class background of respondents. Our discussion concentrated more on the variations in the extension of the life time than on variations in the extension of the present.

In the next chapter we explore further variations in perceptions of the present and the meanings of time associated with them. For, no matter what our conception of the passage of time, be it linear or spatial, we do live in the present and experience directly only those events and sensations occurring in the present. In exploring the perceptions of the present we will keep in mind our earlier discussions of the instantaneous and extended present and how these two perceptions relate to linear and spatial conceptions of time.

NOTES

1. See Mircea Eliade, *Cosmos and History, The Myth of the Eternal Return* (New York: Harper & Row, 1959); see also Viktor E. Frankl, Time and Responsibility. *Existential Psychiatry*, No. 3 (Fall 1966), 361–366. Also in this regard, Tillich wrote: The general tendency is to distinguish existential or immediately experienced time from dialectical timelessness on the one hand, and from the infinite quantitatively measured time of the objective world on the other. That qualitative time is characteristic of personal experience is the general theme of the Existential Philosophy." Paul Tillich, *Theology of Culture* (New York: Oxford University Press 1965), p. 99. Finally, as Kracauer suggested, historians too, have attacked the value of conceiving of time in linear terms even when they attempt to explain events in the context of their occurrence *in time*. Siegfried Kracauer, Time and History. In *History and the Concept of Time* (Middletown, Conn.: Wesleyan University Press, 1966), pp. 65–78.

2. Henri Bergson, *Time and Free Will* (London: Allen and Unwin, 1910), p. 128. In *Matter and Memory*, Bergson wrote: "If I consider duration as a multiplicity of moments bound to each other by a unity which goes through them like a thread, then, however short the chosen duration may be, these moments are unlimited in number. I can suppose them as close together as I please." (New York: Humanities Press, 1970), page 143.

3. *Ibid.*, pp. 130 and 131.

4. What the existentialists, particularly Heidegger, accomplished was a separation in theory of immediately experienced time from calculable time. See Paul Tillich, *op. cit.*, 1964, p. 99. See also R. Plutchik, *The Emotions: Facts, Theories and a New Model* (New York: Random House, 1962).

5. Cohen, Subjective Time. In *The Voices of Time*, J. T. Fraser (Ed.). On a related aspect, see M. L. Farber, Time Perspective and Feeling Tone: A Study in the Perception of the Days. *Journal of Psychology*, **35**, (1953): 253–257.

6. See Fraisse *The Psychology of Time* and M. Wallace and A. I. Rabin, Temporal Experience. *Psychological Bulletin*, **57** (1960): 213–236; see also J. J. Platt, R. Eisenman, and E. DeGross, Birth Order and Sex Differences in Future Time Perspective. *Developmental Psychology* 1 (1969): 70; and J. S. Bruner and C. C. Goodman, Value and Need as Organizing Factors in Perception. *Journal of Abnormal and Social Psychology*, **42**, (1947): 33–44.

7. J. Cohen, C. E. M. Hensel, and J. Sylvester, An Experimental Study of Comparative Judgements of Time. *British Journal of Psychology*, **45** (1954): 108–114. On a related

point, see B. S. Gorman, et al., Linear Representation of Temporal Location and Stevens' Law. *Memory and Cognition*, 1 (1973): 169–171.

8. Sylvia Farnham-Diggory, Self, Future and Time: A Developmental Study of the Concepts of Psychotic, Brain-Damaged and Normal Children. *Monographs of the Society for Research in Child Development*. 31, No. 1 (1966).

9. Cited in M. Rokeach, *The Open and Closed Mind* (New York: Basic Books, 1960).

10. We should note again that the perceived extension of the future on the Lines Test, like its content, may be only projected or imagined. As Broad argued: "The past and present are real, but not the future, which is a non-entity; so called judgments about it are therefore, not judgments at all, for they are neither true nor false." Cited in Cohen, *op. cit.*, p. 264.

    Thus, the intervals imagined and marked off for both the future and the period of the life time stand in contrast with the interval marked for the past, because the past is a known and completed segment of time and can be more accurately estimated. On this point, see Kurt Lewin, Time Perspective and Morale. in *Civilian Morale*, G. Watson (ed.) (Boston: Houghton Mifflin, 1942), pp. 48–70; H. A. Murray, Preparations for the Scaffold of Comprehensive System. In *Psychology: A Study of Science* Vol. III, Sigmund Koch (Ed.) (New York: McGraw-Hill, 1959), 5–54; and Paul F. Wohlford Determinants of Extension of Personal Time. Unpublished Ph.D. dissertation, University of Michigan, 1964.

11. John Updike expressed this very feeling in the following words: "The enormities of cosmic space, the maddening distention of time . . . all this evidence piled on, and I seemed eternally forgotten." *Pigeon Feathers* (Greenwich, Conn.: Fawcett, 1963), p. 177.

12. Philip Merlan, Time Consciousness in Husserl and Heidegger. *Philosophy and Phenomenological Research*, 8, No. 1 (September 1947).

13. See St. Augustine *Confessions*. Tr. by J. F. Pilkington (New York: Dial, 1964).

14. Marking off basic temporal periods such as the future or one's life space does not merely differentiate between personal time—the time one knows by direct experience, memory, or anticipation—and historical time—the time one knows only through projection or conjecture. It also differentiates between the discrete qualities of experience through which we can know the basic temporal periods, such as the personal past and future and the historical past and future. We know the personal past through memory of direct experience, while we know the historical past through reports of history; similarly, we "know" the personal future through anticipations of our future and we "know" the historical future through conjecture.

15. In constructing the Lines Test we defined the specific time segments to be marked off; however, nothing in the instructions demands that these be met. Respondents could draw the personal past, present, and future beyond the boundaries of the segment representing their life time. The direction of time passage, furthermore, is not stipulated in the instructions as being either from left to right or from right to left.

16. Cited in Wohlford, Determinants of Extension of Personal Time, pp. 66–67. In the same context, Kummel wrote, "The fact that (man) can put to himself the problem of the possibility of his duration means at once that time is no longer predetermined and fixed in its form and order." Friedrich Kummel, Time as Succession and the Problem of Duration. In *The Voices of Time*, J. T. Fraser (Ed.) (New York: Braziller, 1966), p. 39. On a related point, see Victor Gioscia, Adolescence, Addiction and Achrony. In *Personality and Social Life*. Robert Endleman (Ed.) (New York: Random House, 1967).

17. Wohlford defined personal time in a very different context as the total array of cognitions at any given moment. See Wohlford, *op. cit.*

18. See, e. g., Mary Sturt, *The Psychology of Time* (New York: Harcourt and Brace, 1925); Jean Piaget, *The Construction of Reality in the Child* (New York: Basic Books, 1954); and The Development of Time Concepts in the Child. In *Psychopathology of Childhood*, Paul H. Hoch and Joseph Zubin (Eds.) (New York: Grune and Stratton, 1955); Fraisse, *The Psychology of Time;* and Klineberg, Structure of the Psychological Future. Unpublished thesis, Harvard University, 1966).

19. Fraisse, *op. cit.*, p. 250.

20. See, e. g., H. A. Witkin, *Psychological Differentiation: Studies of Development* (NewYork: Wiley, 1962); Jerome Kagan and Howard Moss, *Birth to Maturity* (New York: John Wiley and Sons, 1962), and Eleanor Emmons Maccoby and Carol Nagy Jacklin, *The Psychology of Sex Differences* (Stanford: Stanford University Press, 1974).

21. *Ibid.*

22. In this context, see Fraisse, *op. cit.;* and Edmund Bergler and Geza Roheim, Psychology of Time Perception, *Psychoanalytic Quarterly*, 15 (1946): 190–207.

23. Strodtbeck, Family Interaction, Values and Achievement. In *Talent and Society*, D. C. McClelland, et al. (Eds.). (Princeton, New Jersey: Van Nostrand, 1958).

24. Kluckhohn and Strodtbeck, *Variations in Value Orientations.* See also J. W. Vincent and L. E. Tyler, A Study of Adolescent Time Perspectives. *Proceedings*, 73rd Annual Convention of the American Psychological Association, Chicago, Illinois, 1965.

25. On a related point, see E. S. Battle and J. B. Rotter, Children's Feelings of Personal Control as Related to Social Class and Ethnic Group. *Journal of Personality*, 31 (1963): 482–490.

26. Parsons, *op. cit.;* Parsons and R. F. Bales, *Family Socialization and Interaction* (New York: Free Press, 1955).

27. Cited in Seymour Martin Lipset, *The First New Nation* (New York: Basic Books, 1963), p. 98. For a more thorough discussion of this point, see J. B. Bury, *The Idea of Progress* (New York: Dover, 1955).

28. See, for example, Charles McArthur, Personality Differences between Middle and Upper Classes, *Journal of Abnormal and Social Psychology*, 2 (1955), 247–255; Lawrence L. Leshan, Time Orientation and Social Class, *Journal of Abnormal and Social Psychology*, 47 (1952), 589–592. Banfield, too, in his book *The Unheavenly City* (Boston: Little, Brown, 1968) has made a case for the social classes to be distinguished on the basis of their orientations to time. His ideas, however, have been challenged. See, for example, T. J. Cottle and S. L. Klineberg, *The Present of Things Future: Explorations of Time in Human Experience* (New York: Free Press, 1974), Chap. 6.

29. See L. A. Coser and R. L. Coser, Time Perspective and Social Structure. In *Modern Sociology: An Introduction to the Study of Human Interaction*, A. W. Gouldner and H. P. Gouldner (Eds.) (New York: Harcourt, Brace and World, 1963), pp. 638–647.

30. Cf. Edward Tirayakian, The Existential Self and The Person, In *The Self in Social Interaction*, C. Gordan and K. J. Gergen (Eds.) (New York: Wiley 1968). See also Elise E. Lessing, Demographic, Developmental and Personality Correlates of Length of Future Time Perspective. *Journal of Personality*, 36, No. 2 (June 1968): 183–201.

31. Unfortunately, administrative problems precluded collecting comparable data from working- or lower-class students in Vienna.

32. As far as possible test procedures in Austria and America were identical. The time

perception instruments were administered in the identical order in both countries. The same amount of time was given to testing sessions, which were handled in group paper-and-pencil fashion. In Austria *group* refers to schoolroom classes of about 25 people. Instructions were in German and test administration was supervised by teachers.

33. In each case three-way analysis of variance was done always allowing for nonproportional cell representation.

34. Although developmental changes are implied here, to use the term *developmental* would be misleading because a panel study was not done. That is, we have not observed changes in individuals' perceptions over time. Thus, although the term is almost unavoidable, we are in fact dealing with age–group comparisons and not actual developmental transitions.

35. These correlations are presented in Table 31.

36. Lewin, *A Dynamic Theory of Personality*. (New York: McGraw-Hill, 1935.)

37. In examining these results we should again remind ourselves that while our temptation is to speak in developmental terms, we have not in fact performed a developmental study. We are merely comparing one age group with another age group.

39. For a more theoretical statement of this point, see Alex Inkeles, Society, Social Structure, and Child Socialization. In *Socialization and Society*, John Clausen (Ed.) (Boston: Little, Brown, 1968).

39. The ratio between personal, past, and future extensions is, of course, another instance of an estimation based on chronology.

40. On this and associated points, see K. B. Stein, T. K. Sarbin, and J. A. Kulik, Future Time Perspective: Its Relation to the Socialization Process and the Delinquent Role. *Journal of Consulting and Clinical Psychology*, 32 :1968): 257–264.

41. On this point, see Marc Fried, *The World of the Urban Working Class* (Cambridge: Harvard University Press, 1973).

42. The notion of decoupling may provide a better explanation here. Decoupling connotes not merely change but structural disengagement from the past. Thus, college graduation or marriage imply the taking on of new activities with a simultaneous severing of certain prior involvements and associations. Cf. H. L. White and Monro S. Edmundson, *Notes on Coupling and Decoupling*. Unpublished manuscript, Department of Social Relations, Harvard University, 1967. See also Cottle, The Felt Sense of Studentry. *Interchange* 5 (1974), pp. 31–41.

   Looking at the problem from a different perspective, it is ironic that a European writing about American democracy in the nineteenth century should elucidate a central quality of the egocentric perception. Of the Americans he visited in the middle of the nineteenth century, Alexis de Tocqueville wrote: "In the midst of the continual movement that agitates a democratic community, the tie that unites one generation to another is relaxed or broken; every man there readily loses all trace of the ideas of his forefathers or takes no care about them." Alexis de Tocqueville, *Democracy in America*, Vol. II (New York: Knopf, Vintage Books, 1954) p. 4.

43. Cited in Lipset, *op. cit.*

44. On related points, see Lewis Wyndham, *Time and Western Man*. (Boston: Beacon, 1957).

45. W. Lloyd Warner, *Yankee City* Abr. ed. (New Haven: Yale University Press, 1963).

46. One might argue that the block of time in which childbearing is physiologically possible comes to be called the present, because in this discrete time period a woman senses some

greater communion with time. On this point, see Erik H. Erikson, *Identity, Youth and Crisis* (New York: Norton, 1968).

47. Parsons and Bales, *op. cit.*

A discussion of sex-role differentiations is found in this volume in Chapter 6. The maintenance function symbolized by the extended present must be differentiated however, from the so-called maintenance of tradition found in the upper classes. Maintaining tradition, as we noted, would be symbolized by extending the historical past and future rather than the present. A somewhat unintended conceptual implication noted here is that ascriptive or maintenance orientations, at least in terms of the perceptions of time associated with them, tend to emasculate the upper-class man in the sense that they establish a burden of governing the reproduction of generations, a function normally assigned to women.

48. Interestingly, Erikson spoke of the woman's role as mother as ultimately making her home a place where the generations join "in the organizational effort of providing an integrated series of 'average expectable environments.' " Erikson's words support the earlier notion that the middle-class woman's role demands "a minimization of change and a maximization of stability and integrity." See *Identity, Youth and Crisis*, p. 223.

49. As a slightly related point, a hidden dimension of the Lines Test may be the speed of time passage (Chapter 4). Just as space represents extension or length of time, so may it also symbolize the perceived speed of flowing time. Thus, a life space of 70 years is seen not only as a drop in the bucket of all time, it also may be seen as rushing by. Speed should be a significant agent in perceptions of the present as well, where as we have said, chronological guides offer little assistance to the person trying to estimate its linear extension.

50. John Schultz, Custom. In *4x4* (New York: Grove, 1962), pp. 167–168.

51. On this point, see Walter Firey, Conditions for the Realization of Values Remote in Time. In Tiryakian, *Sociological Theory, Values, and Sociocultural Change*.

52. Fred L. Strodtbeck, William Bezdek, and Donald Goldhammer, *Personal Efficacy, Problem Severity and Willingness to Act to Reduce Water Pollution*. Unpublished manuscript, Social Psychology Laboratory, University of Chicago, 1966.

53. On related points, see Orville G. Brim, Jr. and Stanton Wheeler, *Socialization After Childhood: Two Essays* (New York: Wiley, 1966).

54. Some people expressed confusion over the premise of time as a line, which made it difficult for them to designate birth and death marks.

55. Conceptions of this type probably are influenced by religious teachings and affiliations. Indeed, religious philosophies themselves become organized around these very same temporal orientations and questions.

56. The label *egocentric* was not intended to suggest a characterological disturbance or level of personality development, but rather to serve as a focus for defining personal and historical time. Possibly, the egocentric perception represents a misinterpretation of the instructions. Because they end on a note of distinguishing time zones within one's personal life space, the instructions themselves may have distracted respondents from the initial differentiation of personal and historical time. In fact, posttest interviews indicated that some students interpreted the word *time* in the instructions to mean life time, while others thought it meant all time, and still others never did recognize the implication of the interpretation they were being asked to make.

57. Temporal development is contingent upon making the present the intermediate zone- —that is, drawing it larger than one zone and smaller than the other zone.

58. Irrespective of its direction, temporal development is, by definition, automatically destroyed in three ways: The present may be drawn as the largest of the circles, the present may be drawn as the smallest of the three circles, the present may be made equal in size to either or both of the other circles.

59. Approximately 10% of the hospital corpsmen and corpswomen drew present dominance on the Circles Test.

60. We have no real sense of what it means to people to be the age that they are presently. Certainly little in the psychological literature compares with the sociological study of age groups performed by S. N. Eisenstadt. See *From Generation to Generation*, (Glencoe, Ill.: Free Press, 1956).

61. On this point, see Edward Shils, Primordial, Personal, Sacred and Civil Ties. *British Journal of Sociology*, 81)(1957): 130–146.

62. Lawrence Frank, Time Perspectives, *Journal of Social Philosophy*, 4, 1939, 293–312.

63. David Riesman, *The Lonely Crowd* (New Haven: Yale University Press, 1950).

64. Blaise Pascal, *Pensées of Pascal* (New York: Peter Pauper Press, 1946), p. 36. One finds in the passage as well one of the foundations of a concept called Dasein by the existentialist writers.

CHAPTER NINE

DURATION, POTENCY, AND RELATEDNESS:

EXAMINING PERCEPTIONS OF THE PRESENT

BACKGROUND

No matter how one perceives time or accounts for its passage, one can live only in the present. Although we think about yesterday or tomorrow or even the events of only a minute ago, our thinking and experiencing can take place only in the present instant. Yet, although we all may agree that we can live only in the present, people do not agree on their perceptions of the length of the present. The responses of hospital corpsmen and corpswomen, for example, indicated presents ranging in length from seconds to years. While the objective truth is that we live only in the immediate present instant, the subjective truth is that people perceive the present's duration in very different ways. In this chapter we examine how the present is perceived, what the perceptions mean, and how they are related.

We must consider first that the word *present* is used in many contexts. For example, after eating a meal, one might say, "Well, that should hold me for the present," meaning until the next meal. We can speak about the present senior class for the nine-month-long senior year or present world conditions, indicating a period that may be as long as a lifetime. The context in which *present* is used, therefore, can indicate a time duration ranging from the split second in the phrase, "right now at this present instant," to tens or even hundreds of years.

We must also consider that some writers do not think the concept of the present alone is important for understanding the nature of human experience. Alfred North Whitehead, for example, said, "Cut away the future and the present collapses, emptied of its proper content. Immediate existence requires the insertion of the future in the crannies of the present."[1]

Whitehead suggests also that we must have the memory of the past and the expectation of the future to experience the present: "What we perceive as present is the vivid fringe of memory tinged with anticipation."[2] Similarly, the physicist G. J. Whitrow wrote: "Practically we perceive only the past, the pure present being the invisible progress of the past gnawing into the future."[3]

The conceptions of the present are not unique to philosophers and physicists. Let us recall the Navy corpswoman who concluded that as she was perceiving a present experience, it already seemed to be rushing toward the past. She constantly felt as if she were left with nothing but traces of reflections and anticipations. As Whitehead described, she was able to experience the present only by relying on her memories of the past and her anticipations of the future. For this woman, the present by itself was invisible; it became visible only through her memories and expectations.

This perception may not seem strange to us. While we live only in the present, concentrating on the present moment can be difficult, for it seems to move into the past as quickly as we imagine it.[4] The idea of an invisible present goes against the perceptions of the present advanced by St. Augustine: a "present of things past, a present of things present, and a present of things future."[5] Heidegger, speaking from what we would call a linear perspective of time, implied that the past and future as units of time ' are contained in the present. In his words,

" . . . time is conceived as something that is, and its being is characterized as an infinite succession of now points of which only each present now is 'real.' "[6]

The Duration Inventory and the Lines Test helped us to examine variations in estimating the present's duration. The results revealed that people hold different perceptions of the present's duration. Paul Fraisse's assertion that " . . . the present is that which is contemporaneous with my activity. Obviously the changes to which it corresponds are determined by the *scale* [emphasis added] on which I see them"[7] is inadequate because its definition of the present rests on a definition of human activity. Because human activity is as hard to define and describe as is the concept of the present itself, it provides an imprecise basis for defining the present. In the strictest terms human activity is conducted only in the present; past and future activities exist only as recollection and anticipation or imagining. In broader terms imagining is also a form of human activity—in this case mental activity—so existence is the continuous flow of human activity, or one's awareness of his or her activity from birth to death. According to this line of thought, then, the present is actually the period of a lifetime.

One of the problems we face in defining the present is that we recognize that the duration of the present is limited by how accurately our senses measure time. In this regard William James observed, "The moment we pass beyond the very few seconds our consciousness of duration ceases to be an immediate perception and becomes a construction more or less symbolic."[8] In other words, we seem to contrive a definition of the present although we may believe that we are somehow accurately measuring it. Our own test group has shown this to be true. By choosing the units of months and years as well as seconds and minutes to define the present's duration, they have gone well beyond the boundaries of the few seconds of consciousness James mentioned. His words do not resolve the problem of defining the present's duration, but they do help us to clarify two duration issues involved with this problem.

First, one can perceive and define the length of the present by using perceptual stimuli (being pricked with a pin or hearing a bell ring) organized into recognizable units of time (instants, seconds). This kind of perception is objective in the sense that the present is conceived as the instant in which a person touches or hears something. Second, we can define and perceive the length of the present by how we *feel* about external circumstances—for example, working eight hours a day—and the way we organize our perceptions of these external circumstances. That is, we may feel that the work day seems longer or shorter than eight hours. This kind of perception is subjective in the sense that the present can encompass a workday or even a lifetime (as in Fraisse's discussion).[9]

The matter of defining the present remains unsettled (and unsettling). After all, some people may perceive the period of a moment or of a lifetime to be identical in length. In human experience the shifting durations of the present demonstrate that, as Georges Gurvitch said, " . . . there are as many times [or presents] as there are frames of reference . . . there are many times for relativity, to which without doubt, it corresponds . . . but which do not keep absolute duration. Duration is [itself] relative."[10]

To investigate people's perceptions of the present, we first must know how they define where in time the present begins and ends. Then we must know how people perceive the relationship between the present and the past and the present and the future. For example, is the present perceived as a direct continuation of the past, or do people perceive the present and past as overlapping, as we described on the Circle Test. Also, is the future perceived as a continuation of the present, or are the two time zones perceived as overlapping or totally separated from one another. In this chapter we continue our exploration of time perceptions by focusing on how people perceive the present. Specifically, we examine perceptions of its duration, its relationship with the past and future, and people's attitudes toward the present.

A TYPOLOGY FOR THE PRESENT

In Chapter 8 we examined the present's duration in terms of a linear conception of time. Using the Lines Test, we asked our respondents to put four marks on a line indicating where the present begins (presumably, where the past ends), and where the present ends (presumably, where the future begins) and the moments of their birth and death. The space on the line between the marks for birth and death was called life time.

We began our discussion of the Lines Test with the assumption that persons who perceive the present as an insignificant time zone would draw it as a short segment on the time line and persons who perceive the present as a significant time zone would draw it as a long segment on the time line. Thus, respondents could define the present in personal or experiential terms rather than in depersonalized or scientific terms.[11]

The lengths of the present lines, ranging from one-half centimeter to several centimeters on a 20-centimeter test line, were uncorrelated with the lengths drawn for the life time. We used the mean score to divide the presents drawn into groups; we called the shorter segments an instantaneous present and the longer segments an extended present. We suggested that the instantaneous present could be either a symbolic representation of a clock tick, in which case a scientific standard was used to define the present, or a linear representation of an attitude that deemphasizes the present because it is perceived merely as a time zone in which to prepare for the future. We also suggested that the extended present represented a highly subjective response to the present, typically a belief in its general significance, and one need not see the present merely as a bridge between the past and future. Rather, the present may be valued as a distinct time zone.

In looking at the results of the present's duration on the Lines Test, several interesting points about the past and future become clear. For example, if one considers the present to be years long (the extended present), the events of all those years are also labeled *present;* the period considered as the past, then, may well involve the time prior to one's birth. In contrast, regardless of whether one conceptualizes the instantaneous present in chronological terms or as a moment of immediate sensation (what William James called "saddlebacks" of consciousness),[12] we must assume people who perceive an instantaneous present are those who perceive the temporal horizon as a continuous flow of instantaneous presents. Thus, the instantaneous present represents the linear conception of time, while the extended present represents the spatial conception of time. Despite these differences, both the instantaneous and the extended present (so long as it is marked off within the person's life time), can signify a present "existing or happening now; in process, contrasted with past and future," exactly as Webster's dictionary defines the word.

The extended and the instantaneous present can also be considered in terms of *macro-* and *microexperiencing.* In macroexperiencing the present is perceived as extending for the length of a particular activity. When the activity ends, the present ends.[13] One example is the view of the present as the block of time in which childbearing is possible for a person.

In microexperiencing the present lasts only as long as a particular perceptual, muscular, or psychic sensation is perceived to last. For example, one may say that the present is as long as the time during which the doorbell rings. At one extreme, microexperiencing can refer to the smallest amount of time that human beings can perceive,[14] and at the other extreme, to what we experience as a dream—a total, though somewhat extended, moment. In these examples of microexperiencing the person has made an effort to perceive a certain reality, whether it was the reality of the doorbell ringing or the reality of the contents of a dream. This effort contributes to the sense of experiencing the present in microexperiencing.[15] Theoretically, there is no comparable effort expended in the concept of macroexperiencing, for the person is not using an immediately experiencable criterion in defining the present.

THE STUDY

Th distinctions between extended and instantaneous presents are important because they tell us how a person perceives the past, present, and future. For example, if a person perceives the present as a series of moments (instantaneous) he or she should perceive the future as coming after the past and present moments. This person would see change developing out of the natural flow of moments—the natural flow of existence. For this person, the future should contain very few mysteries because it is defined as the continuation of the past and present.[16]

According to this view, one can legitimately contend that the words at the bottom of this page lie in the future. Once they are read, the words fall into the past. Surely this definition of the present and future is a far cry from, say, the future of a year from now, which lies far beyond the activity of reading this page. When the present is perceived in an extended way-—for example, all of this year—the future gets pushed ahead so that it will not arrive until a year from now. In the extended present change is no longer experienced as the logical conclusion of a past and present. The logical conclusion of the present is *still* the present, or more precisely, what the person *calls* the present. "There is nothing else but presents," one corpsman reported in a posttest interview. "Life is either one long present or a series of presents; but it makes little difference, for either way, it's really a stretch of the present. You can only live in the present."

The concept of the extended present develops out of one's institutionalized activities, such as high school years or career tenure. The present becomes the years one is attending school or working on a particular job. The concept of the extended present also implies a present separated from both the past and future. Here the past and future are perceived as lying outside the boundaries of the present. The past and the future are felt to be significant time zones in and of themselves and not merely because they are connected to the present.[17] The future perceived as being disconnected from the present, rather than as its logical conclusion, begins to appear as more mysterious. Are fantasies of knowing the future prevalent among people who perceive an extended present? In the perception of an instantaneous present, the future is the next present in line. Is such a future more easily knowable? Do people who perceive an instantaneous present fantasize about the future?

In addition to investigating the relationship of the instantaneous and extended presents with perceptions of the past and future and with the desire to know the future, we will also study how people feel about the present. Do they, for example, feel that the extended present is especially good or bad or potent? How do people feel about the instantaneous present? If the instantaneous present represents the scientific clock tick, can one really have feelings about it at all? As we explore these questions, we must keep in mind the following notion: The *future* in the instantaneous present perception is the *present* in the extended present perception.

Procedure. The Duration Inventory, the Lines and Circles Tests, the Semantic Differential Scales, and the Money Game have been used to collect data on perceptions of the present.

Although several social scientists have discussed people's orientations to the present,[18] few have ever directly investigated the perceived duration of the present, as we have done with the Lines Test and the Duration Inventory. Our respondents were divided at the median value into two groups: those who drew a long, or extended, present, and those who drew a short, or instantaneous, present.

The Duration Inventory enabled us to study the units of time—for example, hours, minutes, seconds—people select in defining the near and distant pasts and futures, and past and future boundaries of the present. On this test the duration of the present ranged from as short a time as one second to as long as many years or even a lifetime. The results of the Duration Inventory (see Table 3) showed that a majority of hospital corpsmen and corpswomen perceived the boundary between the past and present to be thin; they measured it in clock intervals of seconds, minutes, and hours. However, almost 40% of the men and 50% of the women perceived the boundary between the present and the future as being thick; they measured it in calendar units of days, weeks, months, and years. We called this scheme—thin past-present boundary and thick present-future boundary—the present in advance of itself. Here the past has only recently been encountered and remains close behind the present; the future is postponed by whatever amount of time the respondent chooses to measure the thickness of the present's future boundary.

Respondents who measured the past boundary in years and the future boundary in days drew the scheme we called the present behind itself. They considered the period of right now to be located at the conclusion of the present. In this perception the bulk of the present lies between what the respondent calls the point of right now and the past. Also, the future seems to be brought closer to the present.

Irrespective of whether they perceived the present as being instantaneous or extended, about 60% of the men and 57% of the women exhibited symmetrical presents—the same chronological unit used for the past and future boundaries (see Table 4). About 26% of the men and 34% of the women exhibited a present in advance of itself, while 14% of the men and 9% of the women exhibited a present behind itself.

On the basis of the units chosen by respondents, the Duration Inventory also indicated that the emerging relationship between time zones may be either continuous or discontinuous. The near future may or may not pick up where the present's future boundary left off. In the continuous scheme the near future may begin before the present's future boundary has ended, thereby overlapping the present. In the discontinuous scheme the near future may begin *after* the present's future boundary has ended, thereby leaving a space between the two time zones.[19]

We used the Circles Test to measure the degree to which the present was related to the past and future. We called circle designs *atomistic* when the circles failed to touch, *continuous* when the circles touched at their peripheries only, *integrated* when they overlapped one another, and *projected* when the circles were drawn totally within one another. The distribution of relatedness scores in the Circles Test showed that 60% of the men and 65% of the women drew atomistic designs; 27% of the men and 26% of the women drew continuous designs; and 13% of the men and 9% of the women drew integrated or projected designs (Table 21).[20]

In this chapter we use the Circles Test relatedness scores that refer only to the present's connection with the past and future.[21] We use the results of the Semantic Differential Scales presented in Chapter 6 to examine the degree of potency respondents assigned to the present.[22] Mean scores for this test were computed on the basis of five separate paired adjective scales for the words, *past, present, future,* and *time.*

We consult the results of the Money Game (Chapter 5) to explore people's wish to know the future. The Money Game offered respondents the opportunity to purchase time from the historical past and future and from the personal past and future. The *amount* of money persons offered to recapture the past or know the future was less important than whether they bought any time at all. If people did buy time, we called them fantasizers, a name indicating that they were at least willing to play this game of fantasy. If they refused to buy time, although they had unlimited wealth to do so, we called them realists. Interestingly, the sample of corpsmen and corpswomen was almost perfectly divided, with 47% as realists and 53% as fantasizers.

Results Individual Test Performances and Their Interrelationships. First, we examine the mean scores of all the variables involving the present from the five tests under consideration. These results are presented in Table 39.[23] On the Lines Test American women tended to draw slightly longer presents than American men. However, the lengths of the present drawn by men and women alike seem somewhat long because on the average they occupy about 10% of the line.

The results of the Duration Inventory show that women's scores for both the past and future boundaries of the present were higher than men's scores. Because both of these tests essentially measure perceptions of the present in chronological terms, it is not surprising that the variables are positively correlated showing similar results. However, on the Duration Inventory both men and women demonstrated a present-future boundary that was slightly thicker than the past-present boundary. This suggests the perception of a present in advance of itself.

Looking at the results of the Circles Test, we see that the average perception of the present is nearest to the continuous design (Table 39). Specifically, the male mean score is 2.61; the female mean score is 2.12. To determine the number of people who actually drew atomistic, continuous, and integrated-projected designs, we divided respondents into the three relatedness groups.

Although the sexes exhibit similar potency scores for the present on the Semantic Differential, women's evaluation scores are slightly higher than men's evaluation scores; however, the difference fails to attain a significant statistical level.

Finally, in the Money Game fantasies of knowing the future appeal to men more than to women.[24]

We computed the degree of association between these variables representing various perceptions of the present by intercorrelating the variables. This correlation matrix is found in Table 40.[25] Not surprisingly, the highest correlations obtain between items of a single instrument. The high magnitude of the correlation between the two duration variables suggests a tendency toward perceiving a symmetrical present: The thicker the present's past boundary, the thicker its future boundary. More interesting, only men show a significant relationship between the present's length on the Lines Test and the thickness of the boundaries of the present on the Duration Inventory (average, $r = .22; p < .01$). Although going in the same positive direction, women's values on the same items do not reach statistically significant levels (average, $r = .14$). This difference between the sexes is not so important, however, as the fact that the Lines Test and Duration Inventory are *not* highly correlated. Although both of these tests measure the present in chronological terms, they are tapping different perceptions of the present. While the Duration Inventory asks for the measurement of the present's length in chronological units, the Lines Test inquires into the length of the present relative to the length of one's life time, the historical past and historical future.

The three remaining significant correlations shown in Table 41 all involve women. The first correlation indicates that the longer they draw the present on the Lines Test, the more it is separated from both the past and future on the Circles Test ($r = -.26$). Conversely, the shorter the present is drawn on the Lines Test, the more integrated the present is with both the past and future on the Circles Test.

The second correlation for women is the positive relationship between the present's potency on the Semantic Differential and the thickness of the present's future boundary on the Duration Inventory ($r = .21$).

The less likely respondents were to think of the present merely in terms of scientific or depersonalized clock ticks, the longer they would perceive the present to be on the Lines Test. Thus, the more subjectively a person perceives the present, the more likely he or she will develop specific attitudes toward the present; presumably these attitudes can be measured in part on the Semantic Differential. In examining the figures in Table 41, we discover that a specific attitude—potency—is related *not* to the length of the present on the Lines Test but to the present's future boundary on the Duration Inventory. The thicker the present's future boundary, the more potent the present is judged, and vice versa.[26] We might now speculate that these respondents are hoping that the present as it now is perceived will continue on in the same way in the future.

Finally, high evaluations of the present for women on the Semantic Differential are related to a decreasing interest in knowing the future ($r = -.32$). This is also an interesting finding, given the significant correlations between evaluation and potency ratings of the present (female, $r = .37$) and the thickness of the past and future boundaries of the present (female, $r = .69$). When women evaluate the present as good, they show less interest in knowing the future. Perhaps they feel they already know the future, or maybe their positive attitude toward the present makes them less impatient to know what the future will bring. The future, in other words, can wait its turn. Conversely, a bad rating of the present is correlated with an increased interest in knowing the future. Fantasizing knowledge of the future may be a way of handling one's dissatisfaction with the present.[27]

In summary, for women, high potency ratings of the present relate to a thickening of the present's future boundary, postponing the beginning of the future. High evaluation ratings of the present also tend to discourage the wish to know the future. And, as the length of the present on the Lines Test increases, the present tends to be perceived as being separated from the past and future on the Circles Test.

These last correlations refer only to women. Men demonstrate only one significant relationship among variables of the five different tests—a positive correlation between the duration of the present on the Lines Test and Duration Inventory, a correlation not found in women. The outstanding finding of Table 40 is the low number of significant correlations among the variables. Apparently the five instruments are measuring different perceptions of the present. When one considers, moreover, that all the variables in Table 40, with the exception of knowing the future on the Money Game, pertain directly to the present, the notion of labeling certain individuals, or indeed whole groups of people—for example, working class or lower class—as "present oriented" seems thin, unenlightening and ultimately inaccurate.

The Effect of Present Duration.   Thus far we have examined intercorrelations between the variables involving various perceptions of the present and the fantasy of knowing the future. In this examination we treated the variables in terms of scales—for example, the *number* of centimeters on the Lines Test or the *range* of Duration Inventory responses from seconds to years. Now, we continue our inspection of the results, but instead of using scales, we divide each of the variables into appropriate groups. Present duration scores on the Lines Test are divided into short (instantaneous) and long (extended) groups; the present's boundaries on the Duration Inventory are divided into short (seconds, minutes, hours) and long (days, weeks, months, years) groups, and responses are divided into a symmetrical present, a present in advance of itself, and a present behind itself.

Relatedness scores on the Circles Test are divided into atomistic, continuous and integrated-projected groups; evaluation and potency scores on the Semantic Differential are divided into high and low groups based on mean scores; responses to the Money Game are divided into the groups we have called fantasizers and realists.

Table 41 shows the relationships between the instantaneous and extended present perceptions and all the remaining variables. The significant results are summarized in the following list. The extended present is associated with:

1. For men, thicker boundaries on the Duration Inventory for both the past boundary ($\chi^2 = 15.16$, $p < .001$) and future boundary ($\chi^2 = 19.72$, $p < .000$).

2. Atomicity on the Circles Test for men ($\chi^2 = 14.24$, $p < .001$) and women ($\chi^2 = 14.80$, $p < .001$).

3. For men, high potency ratings on the Semantic Differential ($\chi^2 = 7.85$, $p < .005$);

4. For men, wishing to know the future on the Money Game ($\chi^2 = 5.06$, $p < .02$).

5. An asymmetric present (Duration Inventory), characteristically a present in advance of itself for men ($\chi^2 = 11.62$, $p < .003$) and women ($\chi^2 = 6.99$, $p < .03$).

This set of associations coupled with an extended present represents the spatial conception of time outlined by Bergson. The present is not seen as an instantaneous moment, but as an extended period separated from what went before in the past and what will come in the future. The perceived potency of this extended present period for men does not conflict with, but rather encourages, the wish to know the future's content. However, the extended present is typically a feminine perception.[28]

Looking at the results from the perspective of the instantaneous present, we see an emerging pattern of essentially chronological or linear perceptions of time.

The instantaneous present is associated with the symmetrical present on the Duration Inventory—a present continuous with the past and future on the Circles Test and what we have called the realistic approach to knowing the future on the Money Game. Apparently in the typically masculine perception of an instantaneous present as scientific and somewhat depersonalized, the present is perceived as weak and bad. This is the genuinely linear conception of time. Moment succeeds moment; the past remains irretrievable; the future remains unknowable. The present merely provides the connection between the events and experiences that have just occurred and the events and experiences that are just about to occur.

DISCUSSION

The decision to call a specific stretch of time the present turns out to be a complex one, and a number of perceptions and attitudes go into the making of this decision. A further complication is that women characteristically prefer the extended present, while men prefer the instantaneous present.

One way of trying to clarify the various perceptions of the present we have considered in this chapter is to think of the extended present's effect on perceptions of the future. A person who defines the present in extended terms may want to prolong the present and thus postpone the beginning of the future. Perhaps women in our sample preferred an extended present because the possibility of marriage and childbearing seemed out of line with their present activities in the navy. The women may feel they have more control over their personal outcomes in their present activities than they will have in future activities, especially if their future crucial decisions will be made by husbands.[29] The expectation of becoming a housewife may be more perplexing to the corpswomen than to women who have not worked in the military, which our society considers a masculine institution. Although the sample of corpswomen may seem atypical because of their present work, their performance on almost all the time measures is not significantly different from the performance of the young Austrian women. In addition, the corpswomen do not avoid future experiences on the Experiential Inventory any more than the corpsmen do.

The women's responses to the various instruments in this chapter suggest another interpretation. Because the corpsmen and corpswomen were involved in a 14-week training program, perhaps they perceived the present in terms of these 14 weeks. Such a perception might be influenced by the repetitive activity of the training program, making all the weeks seem identical. This is the notion of public time, in which occupational demands and institutional regulations affect people's perceptions of time.[30] If this is true, one might also argue that on the basis of their preference for an extended present and a spatial or nonchronological perspective of time generally, women may be more sensitive than men to these external or public influences on perceptions of time.[31]

The argument that public and situational factors influence people's perceptions of time, specifically the present, is not entirely satisfactory here because the corpsmen and corpswomen may be influenced by a host of variables. They are at the beginning of both their Navy tenure and their medical training, and because of their youth, they are also at the beginning of their lives. Which factors influence their perceptions of the present is difficult to know, particularly when these factors may yield all sorts of perceptions of the present.[32]

The corpsmen preferred the instantaneous perception of the present, indicating that they are oriented primarily toward the future. Throughout our study, we have suggested that this future orientation,[33] or what

Parsons called instrumental activism,[34] causes men to deemphasize the present's importance. As John E. Smith wrote:

The emphasis on the future, so characteristic of American pragmatism, always points up the importance of a *time for making* or accomplishing something; we do not wait for the future, but anticipate it and seek to determine it.[35]

But not everyone is constantly at work planning for the future. The results presented in this chapter, particularly in Table 42, clearly show that many of the respondents in our study are very much preoccupied with present experiences and events to the extent that they may think only occasionally about the future. For these people, the future may seem to be little more than a new and different present waiting to be encountered. This view of the future is reflected in the perception of the present in advance of itself, a perception exhibited by 26% of the men and 34% of the women in our sample. Here the future is not postponed exactly but is defined within the same chronological period of time other people define as the present.

The perception of an extended present is a complex one. One aspect of this complexity is indicated by the results of the Circles Test, which reveal the following relationship of the present to the past and future. A respondent considers the present to be a period of years running from this precise moment backward toward the past and forward toward the future. But there are also moments prior to this precise moment—the past—but still belonging to what the respondent calls the present, and those expected or anticipated moments—the future—that lie in advance of this precise moment but also within the period of time the respondent calls the present. In effect, this respondent is experiencing two personal pasts and two personal futures. First, there is the past and future that the respondent actually calls past and future—the periods of time lying *outside* the boundaries of the extended present. Then there are the past and present moments that lie *within* the boundaries of the respondent's extended present.

Quite possibly, these two pasts and two futures—what we might call *internal* and *external* pasts and futures—are being reflected in the respondents performances on the Circles Test. The significant association between an extended present on the Lines Test and present atomicity on the Circles Test seems to demonstrate this (see Tables 40 and 41). For the first time, then, we see that separating the present from the past and future on the Circles Test may not actually signify an atomistic conception of the time zones at all. Instead, it may reflect the perception of an extended present, containing its own past and future. By separating the present from the past and future, the respondent shows he or she cares most about the present and less about the past and future. These complexities of perceiving the present are reflected in the words of G. J. Whitrow:

Our mental present must be regarded as the product of an elaborate construction. It is intimately related to our past, since it depends on our immediate memory, but it also determines our attitude to the immediate future.[36]

CONCLUSION

The various tests seemed to tap distinctly different perceptions of the present. Even the Duration Inventory and Lines Test, which we might have expected to be highly correlated, measured different perceptions. More specifically, men and women associated an extended present with a present separated from the past and future. In addition, the more extended the present the thicker its present boundary and the more potent it seems to be.

Where men and women differ is in the finding that men associated an extended present with a desire to know the future (on the Money Game); women did not. Clearly, these findings indicate that perceiving time in a subjective or spatial perspective makes the present a significant time zone in and of itself as opposed to its merely being a bridge connecting the past and future. More important, a subjective perception of time yields a perception of an extended present, a present lasting long enough that one can actually experience it and feel that he or she is living for and in the present. In contrast, a chronological or linear perception of time yields a perception that the present passes so quickly that one can barely experience it. As one goes to find the present, it has already moved on and become the past. In such a linear perspective of time, one can only believe that although instants are tied together in a line, one cannot ever experience an actual instant, but only the flow of instants.

We have discussed several influences on time perceptions—public time, personality, one's role in society, and cultural values. These psychological, social, and cultural influences are external factors. In the next chapter we examine more closely the effect of two such external factors—the achievement value and manifest anxiety—on perceptions of time.

NOTES

1. Alfred North Whitehead, *Adventures of Ideas* (New York: Free Press, 1967), p. 191.

2. Quoted in Whitrow, *The Natural Philosophy of Time*. (New York: Harper and Row, 1961), p. 83.

3. *Ibid*, p. 87.

4. As early as 1882 the psychological literature was introduced to the concept of a specious present. The writer was E. R. Clay in the book *The Alternative: A Study in Psychology* (London, 1882).

5. The Gestalt concept of the unity of organization of mental acts argues that a perceivable present clearly exists. On this point, see Paul Fraisse, *The Psychology of Time*, p. 84. The Gestalt theorist Max Wertheimer built the concept of the unity of organization of mental acts into the law of temporal continuity.

6. Magda King, *Heidegger's Philosophy* (New York: Dell, 1964), pp. 24–25. Whitrow rephrased this idea when he suggested that the present normally has been conceived of as the "coterminus of the past and future." See Whitrow, *op. cit.*, p. 84.

7. Fraisse, *op. cit.*, p. 84. According to Fraisse, variations in the present's extension imply that scale is an essential phenomenological element involved in the sense of a present.

8. *Ibid*, pp. 85–86.

9. The complexity of these issues is evident in the definition of the term extension. Extension normally is defined as an amount of time coming under cognitive consideration. Thus the linear extension of the present reflects a span of awareness or what some writers have thought of as the extension of consciousness. On this point see Paul Wohlford, Extension of Personal Time, Affective States, and Expectation of Personal Death. *Journal of Personality and Social Psychology*, 3 (1966); 559–566, and Richard Cromer, *The Growth of Temporal Reference during the Acquisition of Language*. Unpublished Ph.D. thesis prospectus, Department of Social Relations, Harvard University, 1967).

10. Georges Gurvitch, *The Spectrum of Social Time*, Dordrecht, Holland: D. Reidel, 1964), p. 24. The ambiguities of the present's boundaries are also not resolved by etymological examination. Indeed, the derivation of the word *present* symbolizes its highly subjective character. Derived from the combination of Latin roots *prae* (before) and *esse* (to be), the present is defined in Websters as "existing or happening now; in process: contrasted with past, future." Once again, we are left with the problem of trying to define the time of "now" and "in process."

11. In Erik Erikson's words, the extended present was drawn long enough to be witnessed and judged. See his *Identity, Youth and Crisis* (New York: Norton, 1968).

12. See William James, *Psychology: The Briefer Course* (1892) (New York: Harper Torchbooks, 1961). Also in this context see Daniel W. Hering, The Time Concept and Time Sense Among Cultured and Uncultured Peoples. In *Time and Its Mysteries* (New York: Collier Books, 1962).

13. The present, in other words, suggests a time zone in which one feels a sense of being-in-time. On this point, see Ludwig Binswanger, *Being-in-the-World*. Tr. and with an Introduction by Jacob Needleman (New York: Basic Books, 1963).

14. See Robert Efron, The Duration of the Present. In *Interdisciplinary Perspectives of Time*, Roland Fischer (Consulting Ed.), *Annals of the New York Academy of Sciences*, 138 (1967):713–729.

15. In this regard micro-experiencing resembles the element of activity found in Guyan's

discussion of man's understanding of time. Guyau noted that "the original source of man's idea of time was an accumulation of sensations which produced an internal perspective directed towards the future . . . temporal concepts must be traced back to the feelings of effort and fatigue associated with [man's] movements." Whitrow, *op. cit.*, p. 53. For in effect: ". . . one must not merely act but must be conscious of acting, that is, one must be aware of producing a certain effort." *Ibid.*, p. 82. This particular notion of Guyau's is originally attributed to Henri Bergson. See his *Time and Free Will* (London: Allen and Unwin, 1910).

16. The reader is reminded of the discussion in Chapter 6 involving the Future Prediction Scale.

17. See Fraisse, *op. cit.*

18. See for example, Kluckhohn and Strodtbeck; L. L. Leshan, Time Orientation and Social Class. *Journal of Abnormal and Social Psychology*, **47** (1952): 589–592; and T. D. Graves, *Time Perspective and the Deferred Gratification Pattern in a Tri-Ethnic Community.* Unpublished Ph.D. dissertation, University of Pennsylvania, 1962.

19. T. J. Cottle and Peter Howard, Temporal Extension and Time Zone Bracketing in Indian Adolescents. *Journal of Perceptual and Motor Skills*, **28** (1969):599–612.

20. So few people drew projected designs, that the categories integrated and projected were combined.

21. It should be recalled from Chapter 6 that the relatedness variable could be scored for the entire design or for anyone of the three individual circles.

22. See also Tables 13 and 14.

23. In Table 40 present extension (Lines Test) is measured in centimeters; duration boundaries (Duration Inventory) are measured in clock and calendar units; present relatedness (Circles Test) is measured on an artificial continuum increasing by intervals of two; semantic differentials use scales ranging in units from 1 to 7; and preknowledge fantasies (Money Game) are measured in dollar intervals ranging from $0 to $10,000.

24. Men and women alike adjust their investment according to the allotment of time offered them: The greater the amount of time, the more money they offer.

25. In the computation of the intercorrelation matrix shown in Table 41, the three allotments (hour, day, year), of the Money Game were summed and labeled, simply, preknowledge. Thus, realists entered the correlation matrix with scores of 0, and fantasizers with scores of 1. This procedure was followed because the three allotments were themselves so highly intercorrelated (average male, $r = .67$; average female, $r = .65$).

26. If strong feelings toward the present are in any way operationalized in the potency variable, this correlation supports theoretical contentions like those of Lewin and Tompkins and an empirical demonstration by Wohlford that positive affect increases temporal extensions, and particularly pretension, which Wohlford defined as "extension into the future." See K. Lewin, Time Perspective and Morale. In *Resolving Social Conflicts. Selected Papers on Group Dynamics*, G. W. Lewin (ed.) (New York: Harper and Bros., 1948), pp. 48–70; Sylvan Tompkins, *Affect-Imagery-Consciousness*, (2 Vols. (New York: Springer, 1962); and Paul Wohlford, Extension of Personal Time, Affective States, and Expectation of Personal Death. *Journal of Personality and Social Psychology*, 3 (1966): 559–566.

27. A discussion of this idea is offered in Chapter 6.

28. Women revealed a more extended present on the Lines Test than men. This difference reaches even greater statistical significance when the sexes are compared on a combined median division ($\chi^2 = 11.26$; $p < .001$).

29. J. W. Vincent, An Investigation of Constricted and Extended Temporal Perspectives. Unpublished Master's thesis, University of Oregon, 1965.

30. See, for example, Wilbert Moore, *Man, Time and Society.* (New York: Wiley, 1963); Paul Fraisse, Of Time and the Worker, *Harvard Business Review,* (May–June 1959): 121–25: and Sebastian de Grazia, *Of Time, Work and Leisure* (New York: Twentieth Century Fund, 1962).

31. See H. A. Witkin, *Psychological Differentiation: Studies of Development* (New York: Wiley, 1962).

32. Bergson argued that the organization of personal experiences has more influence on perceptions of time than any external factors providing the original stimulation of the experience. Later, Ernst Cassirer took this same argument one step further when he wrote: "As opposed to the mere world of things, the pure ego may in a sense withdraw into its absolute solitude and inwardness in order to apprehend and affirm its own original vitality and mobility. It achieves its own form only by forgetting and persistently rejecting all schemata drawn from the world of things." See Cassirer, *The Philosophy of Symbolic Forms,* Vol. III; *The Phenomenology of Knowledge* (New Haven: Yale University Press, 1957), p. 39.

33. E. T. Hall, *The Silent Language* (New York: Doubleday, 1959).

34. Talcott Parsons, Youth in the Context of American Society. In *Daedalus,* Winter, 1962, 91 pp. 97–123.

35. John E. Smith, *The Spirit of American Philosophy* (NewYork: Oxford University Press, 1966), pp. 66–67. Also cited in John Muller, Self, Time and Activity. Unpublished manuscript, Department of Social Relations, Harvard University, 1966, p. 9. Another expression of this dominant orientation to the future and consideration of the present as merely the time of preparation came from Thomas Mann. As his own voyage to America concluded, Mann wrote in his diary: "We go early to bed, tomorrow we rise early. To be ready is all." See Thomas Mann, Voyage with Don Quixote. In *Essays by Thomas Mann.* Tr. by H. T. Lowe-Porter (New York: Knopf, 1957), p. 366.

36. Whitrow, *op. cit.,* p. 82.

CHAPTER TEN

THE AFFECT OF VALUING ACHIEVEMENT AND MANIFEST ANXIETY ON PERCEPTIONS OF TIME

What effect do cultural values, such as the value to achieve, and aspects of personality,[1] for example, the level of personal anxiety, have on perceptions of time? In this chapter we explore these factors.

Previous research has indicated that how one perceives the future can distinguish achieving persons from anxious persons. Investigators like David McClelland[2] have observed a relationship between the need to achieve and an active future orientation. McClelland suggested that achievers, in their desire to prepare for the future, must reluctantly tolerate their present as a period of preparation. Achievers see the present as an unavoidable introduction to the future; the present is a time zone that derives almost its entire worth from outcomes that forever lie in store. While this view of the present may leave achievers with a feeling of incompleteness, McClelland asserted that preparation for the future in the present merges the present and future in a rather powerful way:

It is as if the [achievement] need has served to relate present achievement experiences to future ones, to promote understanding of the present in terms of a wider context. Motives seem to tie the present to the future, the specific to the general and long run.[3]

McClelland's last sentence reminds us of the finding that future-oriented people (as determined by the Experiential Invenotry) perceived time as potent and the present as good (on the Semantic Differential). It suggests that expectations are an important ingredient of achievement, for more than anything else, expectations "tie the present to the future, the specific to the general and long run."

An even more direct linkage between the cultural value to achieve and future orientation was formulated by Fred L. Strodtbeck: "For the modern achiever, there is no legitimate excuse for failure. His sense of personal responsibility for controlling his destiny is enormous."[4] In Strodtbeck's terms, the *learning* process has evolved into a *controlling* process, in which people believe that they can govern their own destinies. People who value achievement do not need to wait for the future; they *make* the future happen, in part by building expectations and intentions. But in the process, these people may feel uneasy if their preoccupations with the future, particularly their expectations, make it seem that they are constantly living beyond their actual present experiences.

This feeling of disengagement from the present is rarely discussed in the literature on valuing achievement; however, McClelland referred to a similar aspect of time perception and achievement: "For individuals with high achievement, time is literally moving faster. They are, so to speak, always a little ahead of themselves."[5] Thus, as the future becomes more significant, the present, as the period of time in which expectations are made, begins to seem far less important.

William Henry described one result of this devaluation of the present as the achievers' chronic sense of the unattained and their belief that always there is someplace to go.[6] In the sense of our discussion, that "someplace" is the future.

People who value achievement want to control their lives,[7] or at least their futures; anxious persons are apprehensive about the future. They may even dread it. People who value achievement believe they should keep changing and find new people and situations.[9] Anxiety seems to increase one's desire not to change, but rather to seek out familiar environments.[10]

The psychological literature reveals that people valuing achievement perceive time very differently from anxious persons. David Epley and David Ricks, for example, demonstrated that high achievers, as measured by their college grades, are more concerned with the future than low achievers.[11] McClelland reported that people who value achievement tend to use the future tense more often than people who do not value achievement when devising stories in response to pictures they are shown.[12] In contrast, anxiety discourages people from thinking about the future, and encourages them to think that personal success is unlikely.

Thus, anxious persons avoid even references to the future in these stories. "The farther into his future a high-anxiety individual roams on a preconscious level," Paul Wohlford wrote, "the more he is frightened by what he sees there, and this causes him to draw back more into the present and past in his conscious thought."[13]

Another significant difference in the way anxious and achieving people perceive time is in the degree of potency each attributes to the future. Achievers do not rate the future particularly potent, but anxious persons rate it highly potent.[14] Here, anxious persons may be confessing their inability to confront the unknown realistically or to cope with their fear of the future. Achievers, however, acknowledge not only the importance of the future, but that most of what they do in the present is directed toward the future.[15] For achievers, the future is the time when present activities logically are concluded.[16] As a result of this perception, they can accept the inevitability of the future more easily than anxious persons. Briefly then, anxiety causes people to exaggerate the future's potential dangers and to give it greater potency,[17] while achievement values encourage people to believe that they can control the future.

Another way to compare the time perceptions of achievers and anxious persons is to compare the way in which they perceive the relationship between the past, present, and future. Achievers, for example, live with a feeling that their goals are yet to be attained. Even the moment of accomplishment for one task reminds the achiever that he or she has other unfinished tasks.[18] This outlook makes achievers eager to conclude the present and discover the future. In their eagerness to enter the future and be done with the present, achievers set their watches a few minutes fast.[19] But recall from our discussion in Chapter 6 that the orientation to the future is not based solely on planning or expectation; it is also based on the belief that the present and future are in some way connected.[20] A person motivated to achievement will connect these time zones in some way. One question we ask in this chapter, then, is: Are those persons who value achievement the same persons who on the Circles Test demonstrate integrated (overlapping circles) and projected designs (one circle within another)?

The achievers' conception of human behavior as essentially orderly[21] is not shared by anxious persons, who are unable to establish connections among the time zones. In many respects anxious persons reveal attitudes reflecting the disunity of time zones. Contemporary psychologists might see this as a symptom of ego disintegration.[22] An extreme example of an inability to perceive a relationship between time zones is found in Ludwig Binswanger's description of schizophrenia:

In schizophrenia . . . there is a falling apart of the three ecstasies of time, so that they are no longer related to each other, i.e., past, present, and future become unrelated to one another. Accompanying this phenomenon are other changes in temporality. The future becomes less important, the past comes to predominate, i.e., to become exceedingly important, and the present becomes a mere time-span. . . . For the normal person, the future is open in that there is planning for the future, future expectations, intentions which are projected into the future, and so on. The future is not open for mentally ill individuals. For them the future appears empty because nothing is projected into the future.[23]

Anxious persons have difficulty in differentiating the past, present, and future. Their anxiety somehow disengages them from time and encourages them to fantasize about the more familiar past, rather than the unknown future. Thus, while the *future* dominates the time perceptions of achievers, the *past* dominates the time perceptions of highly anxious people.[24]

The connection of time zones, particularly the connection between what one does in the present and how it affects the future, reminds us of Robert White's notion of competence—the belief that one can shape the future through present effort and action.[25] Achievers should demonstrate a connection between the present and future on the Circles Test more often than anxious persons. Equally important, persons who believe they can effect personal outcomes or influence their environment,[26] are not so likely to fantasize about time in dealing with their problems. If people honestly believe the future can be controlled, they do not need to fantasize about the past or about familiar people and experiences. The past is relinquished, and the future remains to be discovered. Achievers would be less willing to recapture their past and know the future on the Money Game than would anxious persons.[27]

In testing these notions we must remind ourselves that a study of people's perceptions of time zones and how these zones are connected involves more than understanding the temporal perspective. We are examining a person's sense of himself or herself in the context of the passage of time. If what we have said about the way achievers and anxious people perceive time is true, the connection among time zones on the Circles Test becomes a very important variable. The separation of circles could symbolize a lack of ego integrity or one's belief that his or her development has not been continuous.[28] Separated circles could also symbolize a dreading of the future; by drawing separate circles, a person may be revealing a fear that next week or next year (the future) will bring a totally different world that cannot be controlled or understood. The anxious person is attracted to the past, instead, because everything about it is known. In the past there can be no sudden changes, no unforeseen circumstances.

THE STUDY

In this chapter we are investigating the effect of valuing achievement and anxiety on six previously examined perceptions of time. Our hypotheses for this investigation are:

1. Achievers will show a future orientation on the Experiential Inventory; they will emphasize expectations and deemphasize the significance of past experiences.

2. Achievers will exhibit a thickening of the future boundary of the present on the Duration Inventory.

3. Anxious people will exhibit a past orientation on the Experiential Inventory.

4. Anxious people will exhibit a thickening of the past boundary of the present on the Duration Inventory.

5. People valuing achievement will draw connecting circles for the past, present, and future on the Circles Test. This hypothesis is based on the notion that personal control of the future and the sense of competence allow one to feel that what he or she does now will shape the future.

6. Anxious people will draw unconnected circles (temporal atomicity) on the Circles Test. This scheme reflects the anxious person's general inability to differentiate time zones and also an inability to believe that present activity can shape the future.[29]

7. Anxiety about the fast pace of life and a fear about the future generally will *increase* as anxiety increases. Anxious persons fear the future because there is not enough time for them to complete their activities and reach whatever goals they have set for themselves. Their anxiety increases the sense that time is fleeting and exhaustable.[30]

8. It will *decrease* the more one values achievement. Achievers do not fear the future because they are always at work shaping the future, or at least making the future seem less forbidding by establishing new aspirations and intentions.

9. Anxious persons, fearing the future, will perceive the future as being more potent (on the Semantic Differential) than nonanxious persons.[31] Whether anxious people will perceive the future as good or bad on the Semantic Differential, however, is open to question. Dread of the future, a characteristic of anxiety, might lead anxious persons to feel that the future is bad. But anxious persons might want to hide their dread of the future by rating it good on the Semantic Differential. Accordingly, we offer no hypothesis about the relationship between anxiety and the way the future is evaluated on the Semantic Differential.

10. The future will not be rated potent on the Semantic Differential, per-

haps because it is already being shaped through present activity. However, on the basis of the theories we have reviewed regarding achievement motivation and achievement values, we cannot say whether achievers will rate the future as good or bad on the Semantic Differential. Accordingly, no hypothesis is offered about this one variable.

11. Anxiety will be associated with a desire to recover time from the personal past and a wish to know the future on the Money Game. More generally, anxiety encourages the use of fantasy in dealing with the past, present, and future. Freud observed that anxiety stimulates the desire for one to retrieve time and rearrange or relive this past in hopes of relieving present problems.[32] The fear of the future (or more generally of the unknown) should be alleviated by the fantasy of knowing the future. Thus, the anxious person should be more eager to purchase time from the future on the Money Game than the person who values achievement.[33]

To hypothesize about relationships between valuing achievement and fantasies of recovering the past and knowing the future (Money Game) is more difficult. According to Strodtbeck, achievers deemphasize their involvements with their parents and more generally with the past.[34] On this basis, achievers might separate the past from the present and future on the Circles Test. But it is difficult to predict whether this deemphasis of the past is related to a desire to recover it on the Money Game.

Similar problems exist in making predictions about achievers' desire to know the future. The achievers' orientation to the future and their interest in learning whether their present plans and intentions will work out as they want in the future suggests that achievers would buy some future time on the Money Game. Conversely, and more realistically, *not* knowing the future may help people to keep at the task of trying to shape it. Why work at shaping the future if one already knows how the future will turn out?[35] While anxiety would seem to stimulate the use of fantasy as a way of resolving tensions about the past, present, and future, the reactions of achievers to the fantasies offered in the Money Game are more difficult to anticipate.

The Instruments.   We are introducing three new instruments for our investigation in this chapter—the Achievement Value Inventory (VAch), which measures how much people value achievement,[36] the Taylor Manifest Anxiety Scale (TMAS), in which people report the degree of their own psychosomatic anxiety,[37] and the Temporal Anxiety Scale (TAS), which is an attitude questionnaire dealing primarily with anxiety about time. The TAS was designed during our research.

The Achievement Value Inventory   We administered the Achievement Value Inventory along with the time instruments to learn whether valuing achievement is associated with a future orientation. The Inventory was designed by Strodtbeck on the theory that valuing achievement implies individual and autonomous action, personal control of destiny, and a minimization of dependency on one's parents.[38] The VAch scale, which is reproduced in Table 42, is noteworthy because all seven of the items are related either directly or indirectly to perceptions of time. To subscribe to the achievement value on this Inventory, therefore, means holding specific perceptions of the past, present, and future.

Respondents were instructed to agree or disagree with each of the seven items on the Inventory. One point was earned each time a response conformed to the precoded achievement response (underlined in Table 42), with the maximum score being 7. According to the instrument, to value achievement means to disagree with all seven items. Results of the Achievement Value Inventory indicated that the men and women in our sample value achievement almost equally (male, $\overline{X}$ = 5.6; female, $\overline{X}$ = 5.5). On the basis of these results, we divided the corpsmen and corpswomen into high achievers and low achievers, depending on whether their VAch score fell above or below the mean score.[39]

Taylor Manifest Anxiety Scale.   Manifest anxiety was measured by a shortened form of the instrument, the Taylor Manifest Anxiety Scale (TMAS).[40] The TMAS instructs respondents to indicate—by agreeing or disagreeing with each of 21 statements—whether various symptoms of anxiety pertain to them. The underlined answer, as shown in Table 43, signifies the anxiety response. Unlike the Achievement Value Inventory, the TMAS includes no items requiring time perception. In scoring the instrument respondents received 1 point each time a response to a statement corresponded to the anxiety response. The higher the score, the greater the level of manifest anxiety; maximum score is 21.

The Temporal Anxiety Scale.   The Temporal Anxiety Scale (TAS) is made up of 12 statements, each exploring a different aspect of anxiety about time. The statements were developed from discussions with patients exhibiting extreme anxiety neurosis and, in a few instances, psychosis. Respondents were instructed to agree or disagree with each statement according to the following guidelines:

+1 Slight agreement,
+2 Moderate agreement
+3 Strong agreement

−1 Slight disagreement
−2 Moderate disagreement
−3 Strong disagreement

These values were then transposed to generate a 7-point scoring code: Scores of 1 to 3 represented degrees of disagreement; scores of 5 to 7 represented degrees of agreement; 4 served as a neutral point.

The 12 anxiety items of the Temporal Anxiety Scale were randomly placed within a larger time perception inventory and administered together as a 39-item inventory. The remaining items of the inventory are not considered in this particular study.

A factor analysis was performed on all 39 items. As was the case with the development of the Future Commitment Scale (see Chapter 6), a principle axis solution and Kaiser varimax rotation operation was employed. The results in Table 44 show that all 12 temporal-anxiety items load together on the first factor. The remaining five factors were given names based on the items loading on the respective factors, but we do not consider these five remaining factors here. In studying Table 44 note that all items of the TAS load positively, with the exception of Item 15, "I live in the present." Thus, persons tending to agree with the other 11 statements of the TAS tend to disagree with this one statement, and vice versa.

The problems faced in developing the TAS are almost identical to those faced in developing the Future Commitment Scale. With both instruments we make the assumption that particular items will cluster together on the basis of responses to them. The items are then randomly assigned positions in inventories where the remaining items are not essentially too dissimilar from them. The entire instrument was then administered to a large sample of corpsmen and corpswomen. A factor analytic methodology was used to determine whether in fact the items judged *a priori* as temporal anxiety or in the case of the Future Commitment Scale, personal prediction, clustered together on a single factor.

In the following discussion TAS values refer to mechanically computed factor scores in which the degree of agreement or disagreement (on the 7-point scale) is weighted for each item by the degree to which that particular item loads on the factor. The weighting is indicated by the value of the loading shown in Table 44. Thus, for example, Item 32, "I'm afraid I won't be able to lead a full life," is a more powerful indicator of temporal anxiety than Item 11, "Often I think how nice it would be to stop time." The higher the factor score, the greater the expression of temporal anxiety.

Taken together, the 12 temporal-anxiety items reveal facets of anxiety not often recognized as being closely associated with the way people perceive time. However, Howard Harvitz, an American writer traveling in the Soviet Union in a Bernard Malamud short story, does describe himself in terms that closely resemble the items of the TAS:

I consider myself an anxious man, which, when I try to explain it to myself, means being this minute halfway into the next. I sit still in a hurry, worry uselessly about the future, and carry the burden of an overripe conscience.[41]

Measured Intelligence. In all research on attitudes and perceptions an important factor is how much the respondent's measured intelligence influences the results. The intelligence factor is especially important when the test instruments are untried and the information they seek to elicit is complex. Data from research on the relationship between measured intelligence and perceptions of time has not been consistent; however, research indicates that higher measured intelligence tends to increase the extensions of the future on certain time tests.[42] For example, if respondents are shown a picture and asked to write a story about it, those who score high on intelligence tests tell stories extending further into the future than persons with lower intelligence test scores.[43]

In preceding chapters we have referred to the possible effect of the level of intelligence on perceptions of time. In our own work, however, we discovered that a respondent's measured intelligence has little or no effect on the way he or she performs on the various time perception measures, perhaps because our respondents have a somewhat similar intellectual background and are within the same range of measured intelligence.

Nonetheless, we do examine the possible effects of measured intelligence on time perception, valuing achievement, and degree of anxiety.[44]

Results. Results of testing are shown in Table 45. Men and women performed almost identically on the General Classification Intelligence Test (male, $\overline{X}$ = 113.4; female, $\overline{X}$ = 109.6), Achievement Value Inventory (male, $\overline{X}$ = 5.62; female, $\overline{X}$ = 5.41), Taylor Manifest Anxiety Scale (male, $\overline{X}$ = 8.73; female, $\overline{X}$ = 9.39), and Temporal Anxiety Scale (male, $\overline{X}$ = 15.15; female, $\overline{X}$ = 14.48).[45] This similarity in performance between the sexes is important to remember, for in the following sections we begin to see sex differences.

Temporal Correlates of Valuing Achievement and Manifest Anxiety, and the Effect of Intelligence. Several of our original hypotheses are supported by the correlations shown in Table 46. Valuing achievement was associated with an *increase* in temporal relatedness scores on the Circles Test (male r = .25, p < .01; female, r = .26, p < .05), and a *decrease* in scores on the Temporal Anxiety Scale (male, r = −.28, p < .01; female, r = −.28, p < .05). Second, our hypothesis that the future orientation characteristic of achievers would emerge in a thickening of the future boundary of the present on the Duration Inventory was found only among women (male, r = −.02, n.s.; female, r = .20, p < .05). Third, manifest anxiety, as hypothesized, was associated with an increase in temporal anxiety scores for both sexes (male, r = .55, p < .01; female, r = .48, p < .01), but a decrease in temporal relatedness scores on the Circles Test among men only (male, r = −.26, p < .05; female, = .11, n.s.). Fourth, the hypothesis that anxiety encourages people to use fantasy as a way of dealing with the past, present, and future is only partially confirmed. Only men showed a positive correlation between recovery of past time on the Money Game and the Taylor Manifest Anxiety Scale (male r = .26, p < .05; female r = .12, n.s.). Measured intelligence is significantly correlated with temporal relatedness on the Circles Test (Table 46). The higher a person's measured intelligence, the more likely he or she is to connect the circles (male r = .21, p < .01; female r = .20, p < .05).

Although measured intelligence seems to have little relation to performance on the other time perception variables, it is positively correlated with valuing achievement (male r = .24, p < .01; female r = .23, p < .05) but remains uncorrelated with manifest anxiety (male $r = -.04$, n.s.; female $r = -.07$, n.s.). Table 47 also reveals that valuing achievement is negatively correlated with manifest anxiety, although the correlation reaches a statistically significant level only for women (male $r = -.14$, n.s.; female $r = -.22$, $p = <.05$).

We divided our sample into high and low intelligence groups to test the remaining hypotheses and to continue our examination of the effect of measured intelligence on perceptions of time. The dividing point was the combined male and female mean score. We then divided the sample into high and low achievement-value groups in the same way. The range of Taylor Manifest Anxiety scores was so great that we decided to divide the male sample into high-, medium-, and low-anxiety groups and the smaller female sample into high- and low-anxiety groups. The women were divided at the mean score.

The time perception variables—Semantic Differential Scores, Duration Inventory values, and Temporal Anxiety scores—were all divided at their respective combined male and female means into high and low groups. Temporal relatedness designs on the Circles Test were assigned to one of three typologies—atomicity, continuity, or integrated-projected. On the Experiential Inventory those persons listing 3 or more future experiences were called future-oriented, those listing no future experiences were called future-avoidant, and those listing one or two future experiences were called middle. Finally, on the Money Game respondents were divided into fantasizers, who were willing to play the fantasy games of recovering the past or knowing the future, and realists, who were unwilling to do so.

We hypothesized that achievers would exhibit future orientations on the Experiential Inventory and would not rate the future as potent on the Semantic Differential. While both hypotheses tend to be supported by the data presented in Table 48, the finding regarding the future orientation of achievers is statistically significant only among men in the *high* intelligence group ($\chi^2 = 6.43, p < .05$). Achievers' low ratings of the future on the Semantic Differential are statistically significant only among men in the *low* intelligence group ($\chi^2 = 4.58, p < .05$). An additional finding shown in Table 48 is that high measured intelligence raises the percentage of men in the future-avoidant group (35%), but lowers the percentage of men in both the future-oriented (36%) and middle groups (29%).

Among women, valuing achievement tends to be associated with thick past and future boundaries of the present on the Duration Inventory, but this association reaches a statistically significant level only among the high intelligence group ($\chi^2 = 4.70, p < .05$), Furthermore, the positive association of temporal relatedness on the Circles Test with valuing achievement remains independent of intelligence ($\chi^2 = 4.99, p < .10$). Essentially, it was the high achieving woman who exhibited integrated and projected designs on the Circles Test.[46]

The results involving manifest anxiety (Table 49) show that moderate and high anxiety were associated with temporal atomicity on the Circles Test ($\chi^2 = 13.68, p < .01$) among men in the low intelligence group. Among men in the high-intelligence group, moderate and high anxiety were associated with high evaluations of the future on the Semantic Differential ($\chi^2 = 5.68, p < .10$). They were also highly associated with the fantasy of recovering the past ($\chi^2 = 8.32, p < .05$) and knowing the future ($\chi^2 = 5.39$, $p < .05$) on the Money Game. For men and women alike, temporal anxiety was related to manifest psychosomatic anxiety (TMAS) in both high- and low-intelligence groups (low-intelligence males, $\chi^2 = 11.09, p < .01$; high-intelligence males, $\chi^2 = 26.84, p < .01$; low-intelligence females, $\chi^2 = 2.99$, n.s.; high-intelligence females, $\chi^2 = 6.24, p < .01$). These results too are presented in Table 49.[47]

DISCUSSION

Our inquiry into the relationships of achievement values, manifest anxiety, measured intelligence and perceptions of time confirmed most of our hypotheses. Hypotheses 3, 4, and 9 were unconfirmed because manifest anxiety was not associated with past orientations on the Experiential Inventory, thick past boundaries of the present on the Duration Inventory, or increased ratings of the future's potency on the Semantic Differential. Our most surprising finding is the number of sex differences; less than one-third of the relationships between VAch and TMAS and the time perception variables are the same for men as they are for women. This discovery is particularly surprising because men and women performed almost identically on all the instruments examined in this chapter.[48]

We found an important clue for understanding the way achievers and anxious persons react to the future in our finding that manifest anxiety tends to increase temporal anxiety, and valuing achievement decreases it. The achiever seems to accept the idea of the unknown and to regard the shaping of the future as a challenge; but the anxious person seems to turn away from the future and prefers to be involved with the familiar people, events, and symbols of the past. Ignoring, for the moment, the influence of intelligence scores, these findings suggest that the anxious person is someone who wants to retrieve the past and foresee the future. The present seems to be disassociated from the past and future. Time is rushing by too swiftly, and the anxious person seems eager to slow down its passage. William Barrett describes the feeling of the anxious person: "Like a man on shipboard looking out over the storm and the foaming wake, he has turned his back on the future and sees time as a wave rushing backward into the past."[49]

This issue of turning back to the past or looking forward to the future is reminiscent of many of the psychological theories and philosophies of time we have reviewed throughout our study. The achiever could be described in terms of Heidegger's authentic man as one "whose basic possibilities have been disclosed to him including the possibility of death, and who stands resolutely up to those possibilities.[50] Freud, in contrast, observed that anxious people feel helpless in the face of the future.[51]

Melges and Fourgerousse argue, "If an individual has no plans or cannot deal with the present in terms of the past, then behavioral disorganization will result."[52] This view may help to explain why anxious persons tend to draw atomistic time zones on the Circles Test.[53] Achievers' belief in their own capacity to shape the future through present activities in a sense characterizes autonomous action. Here, connections between the past, present, and future are perceived, because essentially people feel that they have shaped the present through past activity and will shape the future through present activities. Thus, prior achievement and accomplishments build the belief that planning and working will produce all sorts of possibilities in the future. Again, according to McClelland, "the (achievement) need has served to relate present achievement experiences to future ones."[54]

In philosophical terms, a person first encounters the sense of future possibilities through the act of planning or intending. In working out these plans, following them, and having them realized, one learns that to a certain degree, the future can be shaped by one's own efforts. The concept of planning, however, refers to something more than merely a plan of action for the next day or two, or even year or two. As William Barrett wrote, "not only do we plan specific projects for today or tomorrow, but our life as a whole is a project in the sense that we are perpetually thrown-ahead-of-ourselves-toward-the-future."[55]

This statement is the key to understanding the characteristic perceptions of time held by achieving and anxious persons. Because time moves forward only, we experience this movement as the sense of being future-oriented, of being-ahead-of-ourselves,[56] as a forward look,[57] or what Binswanger called a future ecstasy.[58] As Friedrich Kummel wrote, "Human activity must be future directed."[59] Yet being future-oriented and living with the sense of being-ahead-of-oneself are slightly different experiences for men than they are for women. When men and women learn sex-role patterns and behaviors, they also learn different attitudes toward, and perceptions of, time. The results discussed in this chapter indicate that male achievers list more future experiences on the Experiential Inventory and perceive a greater potency of the future on the Semantic Differential. For women, valuing achievement does not affect perceptions of the future as much as it affects the thickness of the future boundary of the *present* on the Duration Inventory: The more women value achievement, the thicker they perceive the present's future boundary.[60]

These data tend to support the notion that achievement-oriented men work to shape the future and consider this work a constant life goal.[61] These men seem to want to keep the future out in front of themselves. Achieving women work just as hard at shaping the future, but then they attempt to incorporate the future into what they perceive as an extended present.

In contrast, anxious persons fight against the future orientation. They try to act as though they can actually retrieve the past and know the future. This may be because of the feeling described by Freud that one cannot keep control of oneself in the face of dangerous situations.[62] But one would think that anxious persons would grow continuously more unhappy because their fantasies about time cannot ever come true. The past is never retrieved; the future is never known until it arrives in the present, and one's own lifetime is finite. If the past did not determine the present and, in the process, become irretrievable and if birth or the commencement of subjective experience could be known, we could be completely responsible (Heidegger's term) for our actions and being. Yet this is impossible, or rather, man is unwilling to assume such responsibility, as he disavows the sentence of finitude. As Merlan wrote:

It is, we said, this forgetting which assures us of our not being responsible for what we are. To experience our finiteness from before (as our finiteness) in the adequate way would imply the assumption of responsibility for what we are.[63]

Perhaps we have overemphasized the notion that achievement and anxiety yield opposite perceptions of time. The following quote by

Cartwright is a deeply personal expression about the future, containing attitudes that have been associated with both valuing achievement and manifesting anxiety.

I am prepared for the unexpected. I lunge toward it with hunger, even though I've seen so much I seldom encounter a surprise; however, I am most near death when I begin believing I've seen it all. So I continuously doubt the future. My faith is behind me, not out in front. . . . I am pushing at crashing speeds into the unknowns. . . . My power is in me, in all of us. Life is that power.[64]

But the power consists of utilizing the past and present to shape the future and not turning to the past as a way of avoiding what one presently is experiencing and what one anticipates experiencing. In the end, the power of life derives from accepting the reality of time, the reality that one lives only so long.[65]

CONCLUSION

Philosophers and psychologists have done an enormous amount of work on the meaning of time, particularly the phenomenologists who stress "the systematic description of what is given in experience." However, research on the specific perceptions of the past, present and future is still needed. If we are ever to understand the nature and meaning of an orientation to the past, present or future, we must begin with an understanding of how people perceive these time zones.[67] Can we even speak, for example, about a future orientation if for some people the future begins seconds from now, while for other people it begins years from now? The chronological definitions of time zones, as expressed in the Duration Inventory and Lines Test, must be examined more extensively. Also, additional investigations must be made of time perceptions associated with valuing achievement and manifest anxiety, because both of these variables contribute to our understanding of people's attitudes toward the past, present, and future.

Our data provide evidence of significant variations in perceptions of time, but only in one small area of study. The instruments we have used and the variables they have generated may be somewhat simplistic, and our findings and interpretations of them are suggestive at best. But they do form a basis for future research. One important approach for further research is to keep refining research tools and terms, and to improve research strategies and designs, which should produce more precise studies of time perception.[68] A second approach is to encourage imaginative and intuitive investigations for interpreting and explaining research data,[69] which is especially difficult because we are dealing with such highly subjective perceptions and attitudes.

Naturally, the ideal goal for all research is a combination of these approaches. Alone, each is insufficient. One of the many problems we presently face in the social sciences is that our methodological skills often surpass our imagination. Indeed, our analyses of data often seem to be presented in a way that suggests that the methods we use in collecting our data interest us more than the interpretations we make of the data. But when we are working with perceptions of something as rich and as complex as time, no amount of sophisticated analysis of data from even the finest social scientific instruments will be enough.[70] Creative interpretation and imagination must transform this analysis into a genuine understanding of the best explanation for time perceptions.

NOTES

1. On this point, see L. P. Campos, Relationship Between Time Estimation and Retentive Personality Traits. *Perceptual and Motor Skills,*23 (1966): 59–62; Juanita Chambers, Maternal Deprivation and the Concept of Time in Children. *American Journal of Orthopsychiatry*, 31 (1961): 406–416.

2. See, especially, David C. McClelland, *The Achieving Society* (Princeton, N.J.: Van Nostrand, 1961).

3. David C. McClelland, *Personality* (New York: Holt, Rinehart and Winston, 1951), p. 486.

4. Strodtbeck, Family Interaction, Values and Achievement. In *Talent and Society*, D. C. McClelland, et al. (Eds.). (Princeton, N.J.: Van Nostrand, 1958).

5. McClelland, *Achieving Society*, p. 327.

6. William Henry, The Business Executive: *The Psychodynamics of a Social Role. American Journal of Sociology*, 54 (1949): 286–291.

7. Strodtbeck, *op. cit.;* and McClelland, *Achieving Society.*

8. Harold Persky et al., *Life Situations, Emotions and Excretions of Hippuric Acid in Anxiety States.* Research publication, Association of Nervous and Mental Diseases, 29 (1950): 397–306.

9. Marian R. Winterbottom, The Relation of Need for Achievement to Learning Experiences in Independence and Mastery. In *Motives in Fantasy, Action and Society*, J. W. Atkinson (Ed.) (Princeton, N.J.: Van Nostrand, 1958), pp. 453–478.

10. Seeking out familiar environments not only goes against the psychology of valuing achievement, it contributes to the psychological rigidity experienced in the face of shifting learning situations.

11. David Epley and David F. Ricks, Foresight and Hindsight in the T.A.T. *Journal of Projective Techniques*, 27, No. 1 (1963): 51–59.

12. McClelland, *Achieving Society.*

13. Wohlford, Determinants of Extension of Personal Time. Unpublished Ph.D. dissertation, Duke University, 1964. p. 82. Wohlford's words are reminiscent of Freud's definition of anxiety, which was "the reaction to a situation of danger; [usually] circumvented by the ego's doing something to avoid the situation or retreat from it." Sigmund Freud, *The Problem of Anxiety* (New York: Norton, 1936), p. 65.

14. Donald W. MacKinnon, A Topological Analysis of Anxiety. *Character and Personality,* 12 (1944): 163–176.

15. Strodtbeck, *op. cit.*; see also Henry, *op. cit.*

16. Max Weber, The Protestant Ethic and the Spirit of Capitalism. New York: Scribner, 1958.

17. On this point, see William McDougall, *Outline of Psychology* (New York: Scribner, 1923).

18. McClelland, *Achieving Society.*

19. J. B. Cortés, The Achievement Motive in the Spanish Economy between the 13th and 18th Centuries. *Economic Development Cultural Change,* 9 (1960): 144–163.

20. On this point, see J. E. Rychlak, Manifest Anxiety as Reflecting Commitment to the Psychological Present at the Expense of Cognitive Futurity. *Journal of Consulting and Clinical Psychology,* 38 (1972): 70–79.

21. Robert H. Knapp and Helen B. Green, The Judgment of Music-Filled Intervals and Achievement. *Journal of Social Psychology,* 54 (1961): 263–269. On a related point, see T. Pettit, Anality and Time. *Journal of Consulting and Clinical Psychology,* 33 (1969): 170–174; and B. S. Gorman and B. Katz, Temporal Orientation and Anality. *Proceedings 79th Annual Convention of the American Psychological Association, Washington, D.C.* (1971), pp. 367–368.

22. The perception of temporal disunity is reflected as well in Soren Kierkegaard's reference to man's fundamental disunity with himself. See, for example, Kierkegaard, *The Concept of Dread.* Trans. by Walter Lowrie (Princeton, N.J.: Princeton University Press, 1944). (Originally published in Danish, 1844); and Rollo May, *The Meaning of Anxiety* (New York: Ronald, 1950).

23. Cited in Joel R. Kaplan, Ludwig Binswanger's Existential Analysis. *Existential Psychiatry,* 6 (Summer 1967): 247.

24. Frederick Towne Melges and Carl Edward Fourgerousse, Jr., Time Sense, Emotions, and Acute Mental Illness. *Journal of Psychiatric Research,* 4, No. 2 (November 1966): 127–139.

25. See John Muller, Self, Time and Activity. Unpublished manuscript, Harvard University Department of Social Relations, 1967; see also Robert W. White, Motivation Reconsidered: The Concept of Competence. *Psychological Review,* 66 (1959): 297–333.

26. David Bakan, *The Duality of Human Existence.* (Chicago: Rand McNally, 1966). *op. cit.*; and Robert W. White, *Lives in Progress* (New York: Holt, Rinehart and Winston, 1966).

27. This feeling of efficacy in one's approach to the problems of time was described by William James: "Sustaining, persevering, striving, paying with effort as we go, hanging on and finally achieving our intention—this is pure action, this is effectuation in the only shape in which, by a pure-experience philosophy, the whereabouts of it anywhere can be discussed . . . here is causality at work," James, *Essays in Radical Empiricism* (New York: Longmans, Green, 1912), pp. 183–184; also cited in Muller, *op. cit.*, p. 9.

28. On this point, see Erik Erikson, *Identity: Youth and Crisis* (New York: Norton, 1968).

29. Melges and Fourgerousse, *op. cit.*; see also R. W. White, *op. cit.*

30. See Melges and Fourgerousse, *op. cit.* This notion holds true in the context of feeling states as well as in the performance of anxious persons performing psychophysical estimation studies. Studies have shown that normal persons estimate the passage of clock time more accurately than do anxious persons, who typically perceive time as passing faster than it is chronologically.

31. MacKinnon, *op. cit.*

32. Freud, *op. cit.*

33. Here we make the assumption that performance on the Money Game is not based on purely conscious choice, but that unconscious forces also contribute to the performance on this instrument. Quite possibly, this statement holds true for all the time instruments, but in the Money Game the explicit instructions to fantasize about time might stimulate unconscious factors to a greater extent than is the case with the other instruments.

34. Strodtbeck, Family Interaction, Values and Achievement, *op. cit.*

35. In this regard, see George Herbert Mead, *Mind, Self and Society.* (Chicago: University of Chicago Press, 1934).

36. See Strodtbeck, *op. cit.*

37. Janet Taylor, A Personality Scale of Manifest Anxiety. *Journal of Abnormal and Social Psychology,* 48 (1953): 285–290; and Taylor, Theory and Manifest Anxiety. *Psychological Bulletin,* 53 (1956): 303–320.

38. Strodtbeck, *op. cit.*

39. In this instance mean score refers to the combined male and female mean.

40. The shortened form was tested on more than 250 corpsmen and corpswomen who did not take part in the time perception research. The shortened form correlated highly with the complete TMAS (male, $r = .87$; female, $r = .85$).

41. Bernard Malamud, Man in the Drawer. *The Atlantic,* 221, No. 4 (April 1968): 72.

42. See M. Levine, et al., Intelligence and Measures of Inhibition and Time Sense. *Journal of Clinical Psychology,* 15 (1959): 224–226. On a related point, see P. Kahn, Time Orientation Reading Achievement. *Perceptual and Motor Skills,* 21 (1965): 157–158.

43. See Wohlford, Determinants of Extension of Personal Time; and A. Murray, Preparations for the Scaffold of a Comprehensive System. In *Psychology: A Study of Science,* Vol. III, Sigmund Koch (Ed.). (New York: McGraw-Hill, 1959). From a methodological point of view, moreover, studies demonstrating inter-relationships between intelligence, achievement, and anxiety suggest the necessity of evaluating the effect of intelligence on these two important variables. On this point, see David Levin, *The Prediction of Academic Performance* (New York: Russell Sage Foundation, 1965).

44. For this purpose, intelligence was measured by the Army General Classification Test, which consists of verbal and mathematical subtests. Intelligence is operationally defined as the sum of these two subtests, with each test given equal weight. See General Classification Test, Adjutant General's Office, *Psychological Bulletin,* 42 (1945): 760–768.

45. These mean score values on a 12-item scale should not be misleading. They are themselves computed from responses that range on a 6-point scale from strongly agree (1 point) with a particular item, to strongly disagree (6 points). Mean scores and standard

deviations of variables under consideration in this chapter but not shown in Table 45 have been presented earlier in Table 39.

While the sexes do not exhibit significant differences on the TAS and TMAS, women reveal slightly higher TMAS values than men, but slightly lower TAS values than men.

46. High scores in this discussion refer to scores above the mean on a particular instrument.

47. A common characteristic of Tables 48 and 49 is that where insignificant values obtain, the direction of the percentages normally is the same as in the significant cells. Hence, while intelligence may in some cases reduce the association between variables to levels of statistical insignificance, in no instance does intelligence reverse the effects of achievement and anxiety on the various temporal perceptions.

48. See Tables 39 and 45.

49. William Barrett, The Flow of Time. In *The Philosophy of Time* (Garden City, N.Y.: Doubleday, 1967), p. 362.

50. *Ibid.*, p. 361.

51. Freud, *op. cit.* See also S. Thayer, et al., The Relationship between Locus of Control and Temporal Experience. *Journal of Genetic Psychology*, in press, 1975.

52. Melges and Fourgerousse, *op. cit.*, p. 138.

53. In this context of the relationship between anxiety and perceptions of atomistic time zones, Heidegger used the term *spread* of the temporal horizon. See Martin Heidegger, *Being and Time.* (New York: Harper and Row, 1962).

54. McClelland, *Personality, op. cit.*, p. 486.

55. Barrett, *op. cit.*, p. 361

56. *Ibid.*; See also McClelland, *Achieving Society.*

57. Epley and Ricks, *op. cit.*

58. Ludwig Binswanger, The Case of Ellen West. In *Existence*, R. May, E. Angel, and H. F. Ellenberger (Eds.) (New York: Basic Books, 1958).

59. Friedrich Kummel, Time as Succession and the Problem. In *The Voices of Time*, J. T. Fraser (Ed.) (New York: Braziller, 1966).

60. See Table 46.

61. See Bakan, *op. cit.*

62. On this point, see H. H. Krauss, Anxiety: The Dread of a Future Event. *Journal of Individual Psychology*, 23 (1967: 88–93; and H. H. Krauss and R. H. Ruiz, Anxiety and Temporal Perspective. *Journal of Clinical Psychology*, 23 (1967): 454–455. See also R. D. Hare, Psychopathy, Fear Arousal and Anticipated Pain. *Psychological Reports*, 16 (1965): 499–502.

63. Merlan, Time Consciousness in Husserl and Heidegger. *Philosophy and Phenomenological Research*, 8 (1947): 48.

64. L. Cartwright, *To Make a Difference* (New York: Harper & Row, 1967). Quoted in *Time*, June 9, 1967, p. 90.

65. Anxiety was defined by Heidegger as the qualitative experience of one's own finiteness—that is, one's own awareness of an end to being. Anxiety, in fact, is an announcement or notification of finiteness, for suspended in it are the form and figure of impending death.

66. Barrett, *op. cit.*, p. 368.

67. See G. Poulet, *Studies in Human Time.* Tr. by E. Coleman. (New York: Harper & Row, 1956).

68. See Wohlford, *op. cit.*; and Elise Lessing, Developmental Personality and Other Correlates of Length of Future Time Perspective (FTP). Unpublished research report, 4, No. 2 (1967). Institute for Juvenile Research, Chicago.

69. K. R. Eissler, Time and the Mechanism of Isolation. *Psychoanalytic Review*, 39 (1952): 1–22; Lucille Dooley, The Concept of Time in Defense of Ego Integrity. *Psychiatry*, 4 (1941): 13–23; Georges Gurvitch, Social Structure and the Multiplicity of Time. In *Sociological Theory, Values and Sociocultural Change*, Edward Tiryakian (Ed.) (Glencoe, Ill.: Free Press, 1963), pp. 171–184; Murray, *op. cit.*

70. See Pierre Janet, *L'Evolution de la Memoire et de la notion de temps* (Paris: Chahine, 1928). See also Fraisse, and G. M. Clemence, Time and its Measurement. *American Scientist*, 40 (1952): 260–269.

CHAPTER ELEVEN

CONTRASTING MEN'S AND WOMEN'S PERCEPTIONS OF TIME

CONCLUSIONS AND SPECULATIONS

Understanding orientations to a particular time zone is crucial in the study of time. Therefore, in our conclusions about the meaning of our data in terms of men's and women's perceptions of time, we begin with some remarks on the Experiential Inventory. It is the basis for our notions of temporal orientation.

We first asked our respondents—Navy corpsmen and corpswomen—to list the 10 most important experiences of their life. But by posing the question in this manner, we raised several problems. For example, how do respondents interpret the word *important?* The second part of the instructions asked the respondents to locate their experience in one of five time zones. This request may stimulate various forms of reasoning. For example, a person may list "getting married" as one of the 10 experiences. Will he or she place this experience somewhere in the past because the individual is already married, or will it go in the future because he or she expects to get married someday? This reasoning is obvious of course, but consider the implications of the following finding for the notion of a temporal orientation.

A husband and wife, each working separately on the Experiential Inventory, listed the birth of their son Peter as one of their 10 important experiences. The wife located the experience in the near past; she said her near past included time going back as far as three years, and Peter had not yet reached the age of three. Her husband, however, located the birth of his son in the distant future, literally years from now. He explained that since he no longer was involved with the actual birth date of his son, he now looked forward to the exciting years of pleasure and friendship he would give to and receive from his son.

This example raises two problems. First, how do people interpret the word *experience?* Peter's parents made distinctly different interpretations of the "experience" of Peter's birth. Peter's mother apparently substituted the word *event* for *experience*. Such an interpretation would cause past events to dominate the Experiential Inventory, which would account for the fact that most people listed more past experiences than any other type on the Inventory. Peter's father interpreted experience to mean the anticipation of Peter growing up and maintaining an adult relationship with him.

Second, regardless of whether persons interpret the instructions to mean event or experience, they usually locate each experience in only one time zone. A single event did not mean a single isolated period of time to Peter's father; it meant "future years."

People working on the Experiential Inventory face a perplexing problem: They must take ongoing experiences, like the everyday contact with a child, and break them up so that the experiences can be placed in one discrete time zone, such as the distant future. When respondents begin listing their experiences, they probably begin to think about time and where to locate their experiences in time even before the instructions ask them to do this. Because the instructions of the Experiential Inventory can be interpreted in diverse ways, however, an experience like Peter's existence can be interpreted as a discrete event of the past—his birth—or as the anticipation of future friendship.

In reviewing Bergson's theory of consciousness, Alfred Schutz wrote:

Attention a la vie, attention to life, is therefore, the basic regulative principle of our conscious life. It defines the realms of our world which is relevant to us; it articulates our continuously flowing stream of thought; it determines the span and function of our memory; it makes us—in our language—either live without our present experiences, directed toward their objects, or turn back in a reflective attitude to our past experiences and ask for their meaning.[1]

This concept of an attention to life helps us to understand how people respond to the instructions of the Experiential Inventory. By choosing the basic experiential categories—for example, past-present-oriented or future-oriented, respondents are applying the regulative "attention to life" to the process of locating their important experiences in time—these categories should also represent people's characteristic ways of ordering experiences.[2] Schutz contends that although reflecting on the past and anticipating or expecting the future are different activities of the mind, they encourage us to conceive of time in spatial rather than linear terms.

Reflection revives or recapitulates experiences so that one becomes reinvolved with the past, and expectation is a rehearsal for future action and gives one the feeling that he or she has some sense of what the future will be like.[3]

These notions are complicated by John Dewey's idea that the future remains forever *empty*, because it can be filled only with fantasy, while by definition, the past includes previously enacted real experiences that are retrievable through memory.[4] A future orientation on the Experiential Inventory is a predisposition in the *present* to prepare or rehearse action which may help make an empty period (the future) seem filled. A past orientation, however, because it implies preoccupation with completed action, helps very little in preparing for the future. Thus, we ask if a future orientation on the Experiential Inventory implies a desire to fill a period of time that, by definition, must always be empty. Does planning for the future or anticipating it give people the feeling that they are filling the future to make it seem less ambiguous, less threatening? Does a future orientation imply a belief in one's capacity to shape the future? Or instead, might a future orientation imply a dissatisfaction with the past and a devaluing of memory and reflection?[5]

Similar kinds of questions might be asked about past orientations, or what we have called a past-present orientation. For example, because the past is over, does it stimulate less perceptual and emotional ambiguity than the future? Does the past attract those who are unwilling to deal with perceptual and emotional ambiguities? Does a past orientation suggest a style of thinking distinct from the style of thinking associated with a future orientation? And what of the present orientation, which presumably includes both reflection of the past and anticipation of the future? What sort of style of thought is at work in the present orientation, and what sorts of elements in a person's life contribute to the development of this style of thought? In this same context we again may ask if a future orientation means that one is attracted to the future, threatened by the past and present, or both? Does a present orientation mean that a person has reconciled his or her attractions to and fears of the past and future?[6]

Based on the experiences listed on the Experiential Inventory, we cannot determine whether an experience located in the present indicates that a person is involved with the present, is avoiding the past and future, or is attracted by the past and future. Theoretically, the type of future orientation resulting from a genuine *attraction* to the future should be different from a future orientation resulting from a desire to *avoid* the past and present. What a person is attracted by or seeking to avoid should also influence the person's perceptions of the past, present, and future. Thus, we might hypothesize that a future orientation held by those attempting to avoid the past is more unrealistic and irrational than the future orientation held by those who are genuinely attracted to the future.[7]

Bergson's notion of the attention to life should remind us that each of the 10 experiences a person lists on the Experiential Inventory tells us something about the way the person feels about time. It is not that an individual suddenly labels the distant future as *the* dominant time zone. Rather, locating experiences in the distant future means that an individual spends a certain amount of his or her time thinking about the future and thereby, in Schutz's words, rehearsing for it.[7] In contrast, that people locate their important experiences in the distant past, means that they spend time thinking about the past. They not only reflect on irretrievable experiences, but remain concerned with their responses to these past experiences.

The present orientation is equally difficult to understand, and results from the Experiential Inventory do not clarify this orientation. None of our respondents listed even the majority of their experiences in the present on the Experiential Inventory. Table 1 shows that what is called a present orientation on the Experiential Inventory results from persons listing experiences in past and future time zones as well as in the present. Respondents who emerged with a present orientation were not preoccupied with present experiences; instead, they had labeled as important a group of past experiences and a group of experiences they expect will occur in the future. Given the test results, we cannot discuss a true present orientation on the Experiential Inventory.

The meaning of a present orientation is more clear perhaps in the responses to the Duration Inventory. The continuum of duration responses ranged from the instantaneous present, a scientific and depersonalized perception, through a slightly longer present in which personal action was partly defined by social aspects of time, such as a day's work, to the longest duration of the present, a duration partly defined by sociological and cultural features such as the tenure of one's work. At this extreme, one could say that the present's duration is as long as a lifetime, or as long as a culture survives.[9] There is also a spiritual sense of the present to which Carl Jung alluded:

The man we call modern, the man who is aware of the immediate present, is by no means the average man. He is rather the man who stands upon a peak, or at the very edge of the world, the abyss of the future before him, above him the heavens, and below him the whole of mankind with a history that disappears in primeval mists.[10]

In perceiving the present as instantaneous, it is difficult to know whether a person actually feels experiences or merely observes them. Given such a narrow and depersonalized definition of the present, can such a person really believe he or she can cause events to happen? And in more extended definitions of the present, in which there is a social or cultural point of reference, does the person shape the events in his or her present? Or do the events shape the person's perception of the present's length?[11] Perhaps by choosing how long they believe the present to be, people are shaping its contents too. This idea is similar to the way we have discussed the belief that one can shape the content of the future by one's efforts in the present.

Essentially, the perception of the present as instantaneous implies that the perceiver is really an *observer* of action. Such a person stands back from his or her own life and watches events pass by. He or she can do no more because the present exists on a moment-to-moment basis. In contrast, perceiving the present as extended implies that a person is the *agent* of action; this person makes events happen.[12] He or she does not observe the flow of events, but rather the effect he or she has on the shaping of the present. This is the person Michael Dummett had in mind when he spoke of taking responsibility for one's life.[13] In this sense of agency, this sense of taking responsibility, one develops an understanding of personal autonomy and competence.[14] In this sense the extended present represents the expanse of time that a person believes he or she has shaped or will shape. It includes a portion of past time for which one feels responsible as well as a portion of future time that one believes he or she can shape.[15]

The idea of sense of responsibility complicates perceptions of the extended present because a person may have a greater sense of responsibility for one portion of the present. We spoke about degrees of responsibility in discussing the asymmetric present on the Duration Inventory. While one man may be preoccupied with the future portion of what he calls the present, another man may be preoccupied with what he calls the past portion of the present. These preoccupations were discussed in Chapter 4 as an asymmetric present, which could be a present in advance of itself or a present behind itself (see Table 4). That is, people could be preoccupied with what is just about to occur in the present or what has just occurred in the present. But notice that the word *just* refers to an amount of time for which the person presumably feels some sense of responsibility.[16]

In our work in Chapter 9 we isolated two distinct perceptions of the present—extended present and the instantaneous present. The perception of the extended present was associated with the perception of disconnected time zones on the Circle Test; the present perceived as potent on the Semantic Differential; and the desire to know the future as measured by the Money Game. In contrast, the perception of the instantaneous present was associated with the perception of connected time zones; the present was seen as weak, and little interest was shown in knowing the future. Based on these findings, we might argue that the perceived duration of the present tells us something about the way people adapt themselves to time. How long a person perceives the length of the present indicates the length of his or her awareness—reflective awareness, as David Rapaport called it,[17] or simply, the duration of consciousness.[18]

Presumably, our hospital corpsmen and corpswomen rarely contemplate this duration of conscious awareness when they work on the Lines Test or other time perception instruments. Yet, because the navy environment remains relatively the same for all the corpsmen and corpswomen, they each perceive the length of the present in terms of their own private perceptions developed from selected aspects of their environment. We must recall the public aspects of one's life, such as six hours a day of class and defined times for eating, recreation, and sleeping that influence the time perceptions of navy personnel. When we examine Paul Wohlford's statement that the present's duration may be seen as "the length of the time span encompassed by a cognition,"[19] or according to Rapaport, the "qualitative varieties of reflectiveness inherent in conscious experience,[20] we begin to see how much the environment affects personal perceptions of time, particularly perceptions of the present.

But let us take this notion of the affect of external factors on time perception one step further. Given the similarity of the age and social backgrounds of our respondents, the sex differences in time perceptions suggest that women may be more sensitive than men to the temporal features of their environment—what Wilbert Moore called public time.[21] Men, however, seem to exhibit an orientation to chronological, or moment-to-moment time, which gives them a feeling of time's constancy. No matter what one's mood may be or what demands a particular situation makes on an individual, one can always count on the fact that there are 24 hours in a day, 365 in a year. Women perceive the duration of the present in terms of the social context in which they are living. Their perception seems to be influenced moreover, by a desire to make the present endure.[22] We may speculate, therefore, that men's perceptions of the present's duration are not directly derived from their present experiences.

TEMPORAL CONTINUITY AND DISCONTINUITY

In perceiving the present as extended and delineating the area of the present by the sense of responsibility, people have an understanding of where the past and future connect with the present. But as we have seen, the boundaries between the present and past and the present and future may not be clear. A person may have difficulty saying exactly where the past ends and the present begins. For some people, like those who exhibited atomistic time zones on the Circles Test (Chapter 7) and discontinuous time zones on the Duration Inventory, the temporal horizon is seen as a series of disconnected periods of time.[23] Thinking of time in this manner is much like the feeling one gets when examining a photograph of oneself as a child. The photograph reminds us of a period of time called the past. Although we know it is linked directly to the present, we think of this past as a piece of our childhood that is detached from anything we would now call the present.[24] As strange as it seems, we may feel that 15 years ago is farther away from us in time than is a period 30 years from now. In this view our present activity is directed toward the future, making it seem closer to us.

Let us illustrate the perception of temporal discontinuity by considering the case of three men, all of whom are performing the identical task of preparing for an examination. The first man reports that his preparation for the examination and the examination itself fall within a time period he calls the present. Although the preparation and the actual examination are separated, perhaps by weeks, for him preparation and the examination fall into the same time zone. The second man classifies his preparation period as the present and the examination itself as the near future. He reaches the new present when he actually takes the examination. The period is the same that he had earlier defined as the near future. His sense of continuity in time emerges in part because his activity has involved preparing for a future task. His effort or his personal experiences, rather than his perception of the temporal horizon as continuous, have provided this continuity. For the third man, the present ends literally moments from now, but the future does not begin for months. He sees the preparation for the examination as an ongoing event in the present; the activity of taking the examination, although coming closer with each second, remains "out there" in a time zone he sees as existing between the present and near future. His perception of the preparation and the examination, coupled with his chronological definitions of the temporal horizon, cause this man to perceive a discontinuous temporal horizon. He also believes that he has little or no control over future events. Thus, for him, preparing for the examination and taking it remain seemingly unrelated experiences.

In addition to illustrating the notions of continuous and discontinuous perceptions of time flow, the preceding example also suggests that two forms of reasoning—the *atomistic* and *gestaltist*—can be applied to temporal perceptions.[25] Although these two terms have been defined by many writers, one in particular, Victor Gioscia, has defined them explicitly in temporal terms:

Someone who is accustomed to thinking spatially one part "at a time" is referred to in philosophical language as an "atomist." He is a man who thinks that things consist of a sum total of component parts. . . . He is replaced . . . by the fellow who also thinks in a fundamentally spatial way: He visualizes but takes the entire configuration all at once. We call him a "gestaltist"; things for him do not consist of a sum total of their parts; a thing is a thing, a molecule is a molecule. One can break it up into its component atoms but while one has it, it's a molecule.[26]

The gestaltist notion of time suggests a continuous flow of time from past to present and from present to future. The atomistic notion of time suggests that the time zones are distinct—separated from one another. Apparently, people hold atomistic and gestaltist perceptions of time simultaneously in the same way that they hold linear and spatial conceptions of time simultaneously. For example, the results of the Duration Inventory indicate that the corpswomen and corpsmen perceive the distant past as flowing directly into the near past and the distant future as flowing directly into the near future. The present, however, is regarded separately and does not flow directly out of the near past or into the near future.[27]

The form and instructions of the Duration Inventory may have encouraged people to think about the temporal horizon in atomistic terms. But the Duration Inventory may reinforce the style of thought that causes some people to perceive time in an atomistic manner. If the Duration Inventory does encourage an atomistic style of thinking, one's understanding of causality, the way one event causes another event to happen, which develops in great measure from conceptions of time, is also affected by these atomist and gestaltist styles of thought. Atomists—persons who tend to see parts within a particular whole as being disconnected—should be less capable of perceiving relationships between these parts than gestaltists—persons whose style of thinking urges them to find relationships between parts. In perceiving time, then, the gestaltist sees not only relationships between time zones, but the ways in which events in one time zone affect events in other time zones.

These notions now raise the question of which develops first, a style of reasoning that influences one's perceptions of time or one's perceptions of time, which then help to shape one's styles of thought. Although we do not have sufficient test data to answer this question fully, we might consider the perception of time as discontinuous in light of what Bergson meant by the spatial conception of time. If, for example, the temporal horizon is seen as discontinuous, can people believe in their capacity to affect future outcomes?[28] Does an atomistic perception affect people's conceptions of their own destiny, and, can one even understand the concept of destiny if one perceives the temporal horizon as being atomistic?

The distinction between perceiving the passage of time as being continuous or discontinuous is an important one. Yet both perceptions may suggest that one is so involved with one's own life that one becomes less and less preoccupied with individual time zones or their relationships with each other. We might suggest, then, that on the Lines Test (Chapter 8), the egocentric perception of time (the majority of the time line is delineated as one's life time) would be held by those persons most preoccupied with their own life, personal evolution, and well being. Although the data in Chapter 8 do not wholly confirm this notion, the results of our study suggest that the preoccupation not with one's own life but with the personal future implies an involvement with one's well being and personal accomplishments.[29] When a person is motivated by personal interest or individual achievement, time is defined in terms of how it relates to one's own life, in the form of the egocentric perception on the Lines Test, the experiential future orientation on the Experiential Inventory, and the perception of temporal development on the Circles Test.

The egocentric perception also implies a belief that one's life unfolds as one moves toward the future and that the effect of history plays only a minor role in this unfolding. The egocentric perceiver views the present as an intermediary period of time during which he or she prepares for the future. The egocentric perception stresses making life plans and constantly using the present to prepare for the future. All of this presumably leaves egocentric perceivers wondering how things are going to work out in the future. These people must find a way to evaluate their everyday actions, although these everyday actions are meant to bring rewards later on. Because the egocentric perceiver's life is seen as a series of activities culminating in future rewards never immediately experienced but always in store, egocentric perceivers may begin to experience dissatisfaction with their work and efforts. The image of time unfolding for egocentric perceivers may seem hopeful at first glance, but when one is forever postponing rewards and gratification, life begins to seem incomplete.[30]

The historiocentric perception, in which only a small portion of the time line is delineated as one's life time, implies that one is involved with a period of time shared by many people. It is a time likely to be associated with affiliation needs rather than needs for individual achievement. If egocentrism represents a highly *individualized* perception of time, historiocentrism represents a *collective* perception of time, in that one identifies oneself as being affiliated with a group of people who share a particular period of time. The important aspect of the historiocentric perception, therefore, involves the sharing of time.[31] The egocentric perception implies that 95% of all time belongs to one's personal life time, and the historiocentric perception implies that prior and future generations determine the meaning of one's life. An understanding that history contributes to present identity, perhaps to a greater degree than personal intentions do, is implicit in the historiocentric perception. At very least, this perception is consistent with Erik Erikson's emphasis on "the complementarity of life history and history."[32]

The distinctions between the egocentric and historiocentric perceptions of time and between the instantaneous and extended present, coupled with the idea of mastering one's future, are reminiscent of still one more theory of the meaning of time. In his stages of cultural involvements with time, Oswald Spengler spoke of a Faustian, or modern, epoch in which man perceives time in its infinite extent, but nonetheless, assumes that time can be mastered.[33] Time becomes a force man must harness as he shapes his destiny.[34] In contrast is the Appolonian, or primitive epoch in which time is treated as an aggregate of independent moments, each presumably a present in itself. The past and future have only insignificant meaning. Indeed, they refer mainly to completed and not-yet-arrived presents. We have classified this perception as feminine.

Spengler's distinction between Faustian and Appolonian time hinges essentially on two issues: the recognition of one's finitude—the awareness that one does not live forever—and the belief that one can control time, or at least contribute to the shaping of one's future. Kurt Eissler placed these two issues in the context of psychoanalytic reasoning and asserted that in the Appolonian stage instinct, the *id*, determines the way a person perceives time. In the Faustian stage perceptions of time are determined by the *ego*.[35] A perception of time dominated by the id stresses a belief in immortality and a preoccupation with the immediate present, essentially through impulsive activity. The id demands that impulses be gratified *now*, in the immediate present, without regard to past behavior or future consequences.[36] The ego must control the id and integrate past, present, and expected behavior. The ego, moreover, must be aware that existence is finite. As Lucille Dooley wrote, "To be assured of its unbroken integrity, the ego must feel continuity and activity. The concept of time gives both."[37] The ego initiates and executes the development of such expectations and the perception of the causal relationships between events.[38]

In his clinical observations Eissler isolated certain perceptions of time. The symptoms displayed by several of his patients were a sense of isolation (temporal anxiety), an inability to perceive a continuity between events (temporal atomicity), an intense preoccupation with the present (present dominance), a belief that past events could not affect present events and experiences, and an apparent unconcern for the nature of time's linear flow. One might associate these same symptoms with the perception of the extended present, although the depersonalization Eissler included among the presenting symptoms of these particular patients seems more appropriate to the instantaneous present typology. A confusion occurs here because a moment-to-moment continuity can symbolize the realities of a culture as well as an ego, for like clock time, ". . . the ego [too] conceives of itself as existing in a succession of units of time."[39] Neither the extended nor the instantaneous present, therefore, must be classified as pathological, although the perceptions associated with both typologies may predispose one to various forms of distortions in perceptions of time like those described by Eissler.

We have considered the same perceptions of time as Eissler, but mutual interest does not mean that the corpsmen and corpswomen reveal psychiatric symptons.[40]

REAL TIME AND FANTASY TIME

In compensating or oppositional perceptions of time a person is oriented to the real experiences of the present while fantasizing about recovering past experiences. In one example in Chapter 6 we saw how the future-oriented (on the Experiential Inventory) man wished to recover his past (on the Money Game).[41]

The notion of simultaneous involvement with different portions of the temporal horizon through real experience and fantasy is not new. It has been illustrated in Robert Lifton's observations of Japanese youths. After recounting the dream of one young man, Lifton wrote:

> The dream suggested that [he] was a "bystander" in a more fundamental sense, that he was alienated from those very elements of his personal and cultural past which were at the core of his character structure. . . . Like so many young people in Japan, the youth outwardly condemns many of the symbols of his own cultural heritage, yet inwardly he seeks to recover and restore those symbols so that they might once more be beautiful and psychologically functional.[42]

There are two important points about time in this passage. The first is Lifton's distinction between outward and inward time. As one is outwardly turning his or her thoughts and orientations toward one end of the temporal horizon, he or she may inwardly be involved with the other end of the temporal horizon. Such an outward-inward orientation was displayed in our test results in the example of the past-oriented woman's desire to purchase a year of her future on the Money Game.

The second point involves Lifton's discussion of young people's feelings about their heritage—their historical past. The historical past may seem fascinating because one would like to return to an historical period or because the traditions one presently lives by were established during this time. Lifton suggests that a person preoccupied with his or her own past may also exhibit a special interest in the historical past, particularly in those historical people who have contributed to shaping one's present personality and circumstances.[43] One who denies his or her historical past denies his or her origins. Indeed, many philosophers emphasize the importance of the historical past and historical future by calling it *sacred time;* the time in which one's life is led is called *profane* time. Thomas Becket spoke indirectly about sacred time in these words:

> It is not in time that my death shall be known;
> It is out of time that my decision is taken
> If you call that decision
> To which my whole being gives entire consent.[44]

Sacred time cannot be experienced directly because one lives in profane time, but it can be indirectly experienced by participating in traditional festivals.[45] According to Mircea Eliade, festivals repeat the traditions of the historical past and allow the participants to revive earlier events and have the feeling that they are returning to the historical past.[46] In the distinction made by Eliade between sacred and profane time, we see again the complexities of people's perceptions of time:

The [religious man] experiences intervals of time that are "sacred," that have no part in the temporal duration that precedes and follows them, that have a wholly different structure and origin, for they are of a primordial time, sanctified by the gods and capable of being made present by the festival. This transhuman quality of liturgical time is unaccessible to a nonreligious man. This is as much as to say that, for him, time can present neither break nor mystery; for him, time constitutes man's deepest existential dimension; it is linked to his own life, hence it has a beginning and an end, which is death, the annihilation of his life. However many the temporal rhythms that he experiences, however great their difference in intensity, nonreligious man knows that they always represent a human experience, in which there is not room for any divine presence.[47]

Thus people who experience Eliade's version of sacred time also experience the sense of being able to return to the historical past or to know eternity each moment of their lives.

The passage of time may seem to be metaphysical to some people; other people see it in the form of grains of sand running through an hour glass. In each moment they live, the second group understands that their deaths are drawing closer. John Dewey described such a view this way: "Time is the tooth that gnaws, it is the destroyer; we are born only to die and every day brings us one day nearer death."[48] With each instant, the future flows through the present and collects in the past, and the past remains irretrievable.[49]

As the results of the Duration Inventory suggested (Chapter 4), women's perception of time as an hourglass may be slightly different from men's perception of time as an hourglass. Speaking metaphorically, a present in advance of itself suggests that women look at the passage of time from the perspective of the past end of the present. Unless women look backward from the present, they are in effect looking at the end of the present and not at the past at all. Perhaps this perception keeps these women from "feeling" the large amount of time accumulating in their past.

We might take some people's reluctance to reveal their age as evidence of a denial of an accumulating past. Possibly too it represents a wish to stop time. Clearly, the imperative to stay young is determined by a culture. In particular, men in our culture demand that women appear young because youth and attractiveness are closely associated. From a temporal point of view, men and women alike have to develop a sense of time that keeps them from seeing just how much larger the past is becoming and how much smaller the future seems to be. One can achieve this perspective of time by concentrating strictly on the future or focusing on that precise point of the temporal horizon at which one believes the future becomes the present. There seems to be an endless amount of time at this point on the temporal horizon, and all of it can be used because it is in the present and not in the past. Quite possibly, the perception of a present in advance of itself creates this sense of an endless amount of time accumulating in the present.

THE RELATIONSHIP OF SEX ROLES AND PERCEPTIONS OF TIME

The desire to recover time from the past and to know the future is common to all people. It may be especially intense for persons who seek to deny the past altogether. Surely the desire to recover the past and to deny it altogether affect our attitudes about growing older.[50] At all ages fantasies of recovering the past and knowing the future stimulate the desire to be younger or older than one is presently. To be younger is to be further from death; to be older is to be closer to death. But there are other ramifications of these fantasies as well. For a 14-year-old boy, recovering a year of his past may provide him a retreat from his own sexual development; to an 18-year-old girl, recovering a year prior to puberty may allow her to retreat from womanhood. Similarly, knowing the future may initate fears of the end of sexual capacities as well as fears of death.

One's sexual and sex-role development bear heavily on one's perception of time. Women, for example, know that their childbearing period begins with the onset of menstruation; men experience no clearcut commencement of a period of childbearing years. They learn only that at some point in time sexual potency increases, that at another point it diminishes, and that at still a later point it may cease altogether. Men, therefore, develop a sense of a fearful eventuality awaiting some time in the future.

A man's concern with his career and his more general interest in economic matters may reinforce what sociologists call a market orientation to perceiving time.[51] In the market orientation one must give up or pay something to get something. Applied to perceptions of time, the market orientation encourages men to believe that each act or each moment means one less act or one less moment in his reserve of acts and moments.[52] His belief in a finite amount of goods and services constantly reminds him of the finite amount of time in his life. Women seem to experience the sense of a growing deficit of time less than men. They seem able to take each moment as it comes along without wondering how many moments are left; so counting moments is a less relevant activity for them.

The propositions of masculine exchange and feminine sharing might be relevant for understanding men's and women's performance on the Money Game. The Money Game has been criticized as an invalid instrument because respondents are not asked to part with anything to receive time from the past and future. But in light of the above discussion, the criticism seems biased in favor of masculine ideas about behavior.

We can carry the idea of the effect of the masculine career on perceptions of time one step further. Through his involvement with the occupational and career system, a man learns to evaluate present efforts and conditions essentially in terms of how these efforts and conditions will shape the future. One invests in something now with the hope that in the future one will receive payment of some kind. Psychologists like David Bakan have suggested that the masculine role generally requires that a man actively carve out new situations and possibilities, while the feminine role requires that a woman maintain present conditions and facilitate immediate action.[53]

Social roles, in other words, demand that men learn how to deal with the future and women with the present. The results of the Future Commitment Scale reflect these different roles to some extent, because men are more willing than women to make predictions about global events.

Men do not list more expectations than women or make predictions about themselves more than women. Nor do they bid for knowledge of the future more or less than women. They do not even value achievement more than women. Nonetheless, the relationships between the instruments presented in Chapter 6 suggest that the perceptions of the present and future in fact differentiate the sexes. We can argue, therefore, that the very nature of sociologically determined goal orientations predispose men to treat the present as a series of preparative moments, which they do in our study by expecting or predicting. Instrumental tasks and valuing achievement teach men to accept the uncertainties and ambiguities associated with a future orientation as normal features of everyday behavior.[54]

The results of our study of time perception indicate that men and women have at least three distinct ways of dealing with the future. (None of these ways is entirely rational, because no one can say for certain what the future will bring.) The first two ways of dealing with the future are *intending* and *predicting*. They are based primarily on self-interest and the control of one's fate. Both imply that one believes the future will somehow respond in accordance with one's intentions and predictions, that one is taking some responsibility for the future.[55] The third way of dealing with the future (the least realistic) is to *fantasize* knowledge of it. It may be the most self-indulgent way, and clearly it is the easy way out. By merely wishing one knew what the future holds, one is no longer taking responsibility for the future.

Each way of dealing with the future represents a different method of preparation. An intention promises a certain *content* for the future. One says, "I intend to go to college, and this means that I am planning a great many activities for the future." Predicting the future gives a certain *structure* to the future. By saying, "I predict events A, B, and C will occur," I am giving a particular form to the future. But as we saw in our discussion of the Future Commitment Scale, most of the corpsmen and corpswomen shied away from predicting about themselves. The fantasy of knowing the future says nothing about what one plans to do in the future or even what one thinks will happen in the future. Future fantasy tells only what one *wants* the future to be. A person's way of dealing with the future becomes the way he or she attempts to shape the future through present activities. As George Herbert Mead wrote:

It is true that you can never previse what is going to happen. There is always a difference in what takes place and what has existed in the past. You cannot determine what you are going to be later. But the question is now: What is the relationship of means to ends? We are constantly stating the means.[56]

Someone who perceives a causal relationship between the past, present, and future could be said to perceive time in a *logical* manner. One assumes, after all, that events not only occur in sequence but that this sequence affects the eventual outcome of events.[57] However, someone who draws atomistic schemes on the Circles Tests (Chapter 7) denies not only a connection between time zones, but suggests in this design that there is no causal relationship between the past, present, and future. Such a person could be said to perceive time *illogically*. Both logical and illogical perceptions of time may be the result of particular styles of thought, what psychologists call cognitive styles.[58] However, these different perceptions may also be explained by illustrating the effect of social structure on time perceptions.

Consider, for example, that the women in our study are unmarried, and their present activities in the navy may be drastically changed by marriage. A woman is able to change the direction of her life through marriage; indeed, it may change her perceptions of the past, present, and future as well as her perceptions of the relationships between time zones. Certainly, marriage can immediately alter a woman's present and future, and it may even have an effect on her attitudes toward the past. In some cases it changes a woman because she lives a life that is determined and shaped by her husband. While the status of women and the relationships men and women are establishing, especially in marriage, are changing in American society, the fantasy of the handsome knight who takes the maiden away with him is still alive in certain sectors of American culture. A woman may believe that marriage can bring changes that she cannot presently imagine. In this light a perfectly reasonable arrangement of the circles for women would be one showing the past and present in one plane and the future in the plane of potential change, the plane of the unexpected. Such arrangements did occur on some of the Circles Test designs drawn by corpswomen.

The changes a man undergoes in marriage are quite different. Although men surely redirect their lives and alter their values when they marry, their perceptions of the present and future are not so greatly affected as are women's perceptions of the present and future.

To summarize, marriage, by changing the course of events for women, may affect their perceptions of the causal relationships between events in the past, present, and future. Women who perceive marriage as a whole new start in life believe that their past has had little influence on the shape of their present and future.[59] The fact that most women still change their name in marriage and usually their place of residence may contribute to their sense of disengagement from the past.[60]

This idea that marriage can drastically change one's life and one's time perceptions suggests that important events can disrupt time perceptions. These temporal disruptions can be related to earning rewards through personal effort—achievement—and achieving rewards merely because of social rank or position—ascription.[61] For example, an achievement-oriented marriage would be based primarily on responses to specific performances—"for what one gets out of it"—while an ascriptive oriented marriage would be based primarily on loving one's spouse for their kindness, goodness, and other qualities.

These two orientations to marriage, however, represent two different attitudes. In the strictest sense marriage is an ascriptive interaction that may result in changing a person's sense of how relationships are causally connected. The impoverished woman who marries the wealthy man must deal with a whole new reality and a whole new idea of how her own personal effort may or may not result in certain outcomes.[62] Regardless of how marriage changes one's present and future circumstances, one cannot be rid of the past and the very real events that occurred in the past. No matter what marriage yields, one is always likely to wonder "how things might have turned out if only I had married. . . ." In terms of perceptions of time, we begin to see why people might disconnect circles (atomicity) on the Circles Test and in so doing communicate the belief that there is little or no connection between the past, present, and future.

In Chapter 6 we discussed these two forms of rewards, the ascriptive and the achieved, in terms of what Talcott Parsons called instrumental and expressive interactions. Parsons described the instrumental interaction as one in which people do not expect immediate gratification but, instead, look forward to achieveing later goals, later gratifications. They are concerned with the *ends* of the interaction. In the expressive interaction, people expect the immediate gratification of their needs. They are involved with the *means* of interaction for their own sake.[63] From this work on the means and ends of interaction, Parsons developed the notions of ascription and achievement.

Oversimplifying what Parsons theorized, ascription refers to what one is, while achievement refers to what one has done, is doing, or may do in the future. In defining ascription and achievement Parsons made the issue of time explicit:

Social systems aspects. (1) Ascription: the role-expectation that the role incumbent . . . will accord priority to the objects' *given* attributes . . . over their *actual* or *potential* performances. (2) Achievement: the role expectation that the role incumbent . . . will give priority to the objects' *actual* or *expected* performances and to their attributes only as directly relevant to these performances, over attributes which are essentially independent of the specific performances in question [emphasis added].[64]

The important distinction made by Parsons is between responding to the temporal horizon in terms of what is already *given* in reality, and responding to it in terms of what one might *achieve* through personal effort. We honor the person who achieves riches in one way, the person who is born into riches in another way.

Parsons' distinctions between ascription and achievement suggest not only that people evaluate their own behavior and social objects in terms of historical considerations, but that important events, like birthdays, anniversaries, and marriages, cause changes in people's perceptions of time.[65] At the very least, these ceremonies make us rethink the meaning of the past, present, and future and how, as individuals, we perceive the evolution of our own lives. More generally, the typically masculine instrumental interaction encourages men to perceive the passage of time as continuous. In contrast, the form of typically feminine interaction, which Parsons called expressive, tends to disrupt the perception of time and encourages a perception of time as discontinuous.[66]

No event causes a total disruption in the perception of time, for this would be to assert that one's very sense of identity ends, rather than changes, because of experiences like a marriage or birthday. We might assert, however, that in the developing sense of identity, a sense Erikson describes as a series of accumulating identificatory experiences, the actual identifications may seem more temporally disjunctive for men than for women.

If the above propositions are valid, then individuals undergoing serious changes or indeed persons moving into new cultures should experience serious disruptions in their sense of time. Yet these are instances, perhaps, in which the sense of disruption may be less for a woman than for her husband, since marriage has already prepared a women for experiences of disruption of this sort. By affording women a sense of continuity during a period of uprootedness, marriage in fact may help them to overcome feelings of uprootedness. Men, on the other hand, may experience an intolerable discontinuity in their own sense of time and personal evolution.[67]

CONCLUSION

We have discussed many kinds of time perceptions and offered various interpretations of them. Some of the perceptions may seem highly relevant to the way we live each day, while other perceptions may have little relevance to our lives. Albert Einstein wrote:

The normal adult never bothers his head about space-time problems. Everything there is to be thought about, in his opinion, has already been done in early childhood. I, on the contrary, developed so slowly that I only began to wonder about space and time when I was already grown up. In consequence, I probed deeper into the problem than an ordinary child would have done.[68]

Perhaps the most basic question we can pose about time perception is: Does one's perception of time help a person to understand and appreciate the nature of living and dying?

One of the major features of the ego, Lucille Dooley wrote, is to ward off the anxiety that comes from recognizing one's impermanence. One can view the passage of time in such strict chronological terms that one begins to forget that his or her own life is running out with each passing moment. Or the ego may fragment the passage of time into minute units, each presumably capable of sustaining one's belief that every individual unit can last forever. The expressions "I live only for today" and "I don't think about tomorrow or anything coming to an end" help some people achieve a feeling of permanence by carving out a series of present time zones, with each one so long that a person cannot live long enough to reach the end of any or all of them. The anxiety about the impermanence of life may also be warded off by placing personal experiences into the context of historical ceremonies or the occult.[69] In all these ways we become involved with aspects of time that help us to understand the meaning of a single life and ease the fear of a single life lasting for so short a time.[70]

In the end, these aspects of time perception are of the utmost importance to people as they lead their lives day by day. All human beings must work out their own conceptions of time flow and their own perceptions of the temporal horizon. They must deal with the historical past that existed before their births and with their own pasts, their own presents. They also must deal with their personal future and its unknowable content, just as they must deal with the historical future, a period of time that they will never experience.[71] Finally, all people, influenced by social, cultural, religious, and educational systems,[72] by their own personalities, and by the rules governing their social roles, must attempt to understand and make peace with the reality of their birth and their death.

These are quintessential life concerns that must be at the forefront of any study of time perception. But a study is made even more complicated by the fact that these perceptions will constantly change because people are living in time. In chronological terms the present becomes the past and the future becomes the present; with each second our perceptions of the past, present, and future must be altered. Erikson wrote at the conclusion of his prologue to Identity, Youth and Crisis, that the study of lives must at some level be a study of biographies in history; people do not hold still long enough for anyone to describe them.[73] One implication of this point is that our best effort may be to write only biographies of the dead, because for them time has stopped. But a study of perceptions of time is a study of the movement of living beings in time. People are always in motion, and that very motion is part of the horizon of time that the people in our study have been asked to examine.

All research on human beings must take into account this movement of life and the constant redefinitions of oneself that accompany growth and maturation. An experiment in which the investigator examines before and after conditions is an example of research in which changes and development over time are being studied explicitly. Whether it is mentioned or not, time is a major variable in this type of study. Research like ours, which is aimed directly at perceptions of time, is more complicated because it focuses on what is only the background of many studies. While some researchers may want to know how people have changed from one period of time to another, we want to know how people perceive this period of time and how they perceive the changes in themselves over this period.[74] We want to know how perceptions of the passage of time shape their perceptions of how they may have changed over time.

Our exploration in this study represents only one step in understanding people's perceptions of time. The data we have examined have been generated from previously untried instruments and are at best tentative and suggestive. Our project is made difficult not only by the complexity of time as a variable, but also by the tensions we have noted between the research methods of the natural sciences and the social sciences and the results of employing objective and subjective analyses. The accuracy of measurement and types of systematizing of data that are possible in the natural sciences are not yet possible in the social sciences. While this fact is disturbing to some social scientists—particularly those of us fascinated with time—we will have to be content with the notion that certain human processes, like perceptions of time, do not lend themselves entirely to scientific inquiry. Nonetheless, these qualifications need not constrict our imaginations, and we can make our plans for further research on perceptions of time.

NOTES

1. Alfred Schutz, *The Problems of Social Reality*, Collected Papers; Vol. I (The Hague: Martinus Nijhoff, 1962), pp. 212–13.
2. Once again, one could be argue that the location of experiences in time cannot truly yield temporal orientations because the acts of listing and locating experiences are them-

190 Perceiving Time

selves acts of the present only. Reflecting on past experiences does not carry us back to the past, nor does expectation thrust us forward into the future. As Heidegger said, past and future belong in and to the present, for both are created and filled by activities of the present—memory, anticipation, and so forth.

3. See J. T. Fraser, *Of Time, Passion and Knowledge* (New York: Braziller, 1975).

4. See John Dewey, Time and Individuality. In *Time and its Mysteries* H. Shapley (Ed.) (New York: Collier Books, 1962); and Dewey, *Human Nature and Conduct* (New York: Modern Library, 1957).

5. This notion is similar to Robert Lifton's ideas about Japanese time orientations in which individuals "consult" both ends of the temporal horizon as they consciously and unconsciously turn their attention to locating experiences in time. See Lifton, "Youth and History: Individual Change in Postwar Japan. *Daedalus*, **91** (Winter 1962); 172–197.

6. In this context perhaps only an involvement with actual present experiences should be called a present orientation, because, as Bergson suggested, as any other form of present orientation—for example an equal assortment of past, present, and future experiences—relies too much on recollection and fantasy.

7. Other factors also would determine the eventual orientations to time and the actual contents of people's perceptions of the past, present, and future. Recall, for example, Klineberg's finding that age and stage of personal development affects the perceived form of the future. As people get older, the shape and content of the future tend to become more realistic, leaving fantasies to dominate the future perceptions of young children. How Klineberg's data relate to approach-avoidance dynamics remains another issue, but his research adds to our understanding of how people begin to fill the empty future. See Klineberg, Structure of the Psychological Future. Unpublished thesis, Harvard University, 1965. See also Cottle and Klineberg, *The Present of Things Future.* (New York: Free Press, 1974).

8. Schutz, *op. cit.*

9. See Fraser, *Of Time, Passion and Knowledge op cit.*

10. Carl G. Jung, *Modern Man in Search of a Soul* (New York: Harcourt, Brace, 1933), p. 196. On a related point, see George Kubler, *The Shape of Time* (New Haven: Yale University Press, 1962).

11. This notion is based on Michael Dummett's distinction between the agent and observor of action. See Dummett, Bringing About the Past. In *The Philosophy of Time*, R. M. Gale (Ed.) (Garden City, N.Y.: Doubleday Anchor, 1967).

12. On this point, see David Bakan, *The Duality of Human Existence* (Chicago: Rand McNally, 1966).

13. Dummett, *op. cit.;* see also Richard N. Neibuhr, *The Responsible Self* (New York: Harper & Row, 1963).

14. See Kurt Lewin, *A Dynamic Theory of Personality* (New York: McGraw-Hill, 1935); and Lewin, *Resolving Social Conflicts* (New York: Harper and Bros., 1948); Erikson, Identity and the Life Cycle. *Psychological Issues*, **1**, No. 1 (1959); and Robert W. White, Motivation Reconsisdered: The Concept of Competence. *Psychological Review*, **66** (1959):297–333.

15. On a related point see A. W. Siegman, The Relationship between Future Time Perspective, Time Orientation, and Impulse Control in a Group of Young Offenders and in a Control Group. *Journal of Consulting Psychology*, **25**, (1961): 470–474.

16. Data from the Duration Inventory (Tables 3 and 4) clearly indicate that just-occurred time or just-about-to-occur time can refer to anything from seconds to years.

17. David Rapaport, States of Consciousness: A Psychopathological and Psychodynamic View (1951). In *The Collected Papers of David Rapaport*, Merton H. Gill (Ed.) (New York: Basic Books, 1967), pp. 385–404.

This idea is not too dissimilar from either James' notion of a specious present or the Gestalt theory of mental act organization, except that awareness here is limited by a theory positing a breakdown of constraints as well as "excessively extensive and rigid control of motivations," *Ibid.* p. 391.

18. See William James, *Essays in Radical Empiricism* (London: Longmans Green, 1912). On a related point, see Marie Bonaparte, Time and the Unconscious. *International Journal of Psychoanalysis*, 11 (1940), 427–468.

19. Paul Wohlford, Extension of Personal Time, Affective States, and Expectation of Personal Death. *Journal of Personality and Social Psychology*, 3 (1966: 449–466.

20. Rapaport, *States of Consciousness A Psychopathological and Psychodynamic View*, *op. cit.* p. 400.

21. Moore, *Man, Time, and Society*, *op. cit.* On this point, see also Julius Roth, *Timetables* (Indianapolis: Bobbs-Merrill, 1963).

22. On a related point, see R. P. Kimberly, Rhythmic Patterns in Human Interaction. *Nature*, 228 (1970), 88–90.

23. See Henry A. Murray, Preparations for the Scaffold of a Comprehensive System. In *Psychology: A Study of Science: Formulations of the Person and the Social Context*, Vol. *III* Sigmund Koch (Ed.) (New York: McGraw-Hill, 1959), pp. 5–54.

24. The example of the photograph of childhood illustrates the fact that the concrete nature of already experienced time may seem less real to some people than the time they imagine through their expectations and anticipations. As we saw in Chapters 6 and 10, persons who value achievement tend to dismiss their childhoods and pay special attention to the future. Childhood, therefore, may well seem like a period of time that existed millions of years ago instead of only 10 or 20 years ago.

The example of the photograph is also reminiscent of Erik Erikson's work on identity. See *Identity, Youth and Crisis* (New York: Norton, 1968).

25. See Wolfgang Kohler, *Gestalt Psychology* (New York: Liveright, 1947).

26. Victor Gioscia, Classification of Family Types. Unpublished manuscript, Department of Psychology, Queens College, 1966, p. 3.

27. This conception of the temporal horizon might be illustrated as follows: (distant past- –near past) (present) (near future–distant future) where dashes are meant to represent continuous passage of time and parentheses discontinuous passage of time. On this point, see T. J. Cottle and P. Howard, Temporal Extension and Time Zone Bracketing in Indian Adolescents. *Journal of Perceptual and Motor Skills*, 28 (1969):599–612.

28. Cf. White, *op. cit.*

29. When experiential orientations on the Experiential Inventory and life time designations on the Lines Test are crosstabulated, we see that historiocentrism is not related to a future orientation (male, $\chi^2 = 3.05$, 4 df; female, $\chi^2 = 4.05$, 4 df) but is related in women to a past-present orientation ($\chi^2 = 6.53$, 4 df. $p > .05$).

30. On a related point, see A. P. Mahrer, The Role of Expectancy in Delayed Reinforcement. *Journal of Experimental Psychology*, 52 (1956): 101–105.

31. On this point, see Ervin W. Strauss, Existential Approach to Time. In *Interdisciplinary Perspectives of Time, Annals of the New York Academy of Sciences*, Roland Fischer (Consulting Ed.) 138 (1967): 759–766.

32. Erikson, *Identity, Youth and Crisis*, p. 314. One might also speculate that the historical future, too, shapes personal and social outcomes. Thus, the historiocentric perception, although seemingly illustrating the insignificance of a 70-year lifetime, may also be interpreted as illustrating the psychohistorical viewpoint of a 70-year lifetime. On this point see S. G. F. Brandon, *History, Time and Deity* (Manchester: University Press, 1965).

33. Oswald Spengler, *The Decline of the West* (New York: Knopf, 1926).

34. See also Max Weber, *The Protestant Ethic and the Spirit of Capitalism* Tr. by T. Parsons (New York: Scribner, 1958).

35. K. R. Eissler, Time and the Mechanism of Isolation. *Psychoanalytic Review*, 39 (1952): 1–22.

36. See Bonaparte, *op. cit.*

37. Lucille Dooley, The Concept of Time in Defense of Ego Integrity. *Psychiatry*, 4, (1941): 20.

38. Heinz Hartman, *Essays on Ego Psychology* (New York: International Universities Press, 1964). Spengler's third, intermediate, typology, the Magian (or early Judaeo-Christian), which Eissler correlated with superego time, also is envisioned as infinite in its extent. Unlike the contemporary Faustian man, however, the Magian renders the present a small and transitory moment in a universal temporal order and finds himself, moreover, dominated by the intractable strengths of predestination.

39. Dooley, *op. cit.*, p. 15

40. On this topic, however, see John Cohen, *Psychological Time in Health and Disease* (Springfield, Ill.: Charles C. Thomas, 1967).

41. The notion of compensating perceptions of time is not too dissimilar from Jung's notion of the law of psychological opposites: Where we see a direction and intensity of behavior, we shall find somewhere in that same person the opposite mode of that behavior, again of equal intensity.

42. Lifton, *op. cit.*, pp. 172–197.

43. This idea also is reminiscent of primordial residues and the collective unconscious. If we are motivated by events of our own personal past, might we not also possess residues of the historical past, a period of time we share with others, who are presently alive, as well as with those already dead?

44. Quoted in Maurice Friedman, *To Deny Our Nothingness* (New York: Delacorte, 1967), p. 105.

45. On a related point, see Robert J. Smith, Cultural Differences and the Concept of Time. In *Aging and Leisure*, Robert W. Kleemeier (Ed.) (New York: Oxford University Press, 1961).

46. See M. Eliade, *Cosmos and History: The Myth of the Eternal Return* (New York: Harper & Row, 1959).

47. M. Eliade, *The Sacred and the Profane* (New York: Harper & Row, 1961), p. 7.

48. John Dewey, *op. cit.*, p. 141.

49. For a related perspective in the context of literature, see Jerome Buckley, *The Triumph of Time* (Cambridge: Harvard University Press, 1966).

50. See E. Cumming and W. E. Henry, *Growing Old* (New York: Basic Books, 1961).

51. George C. Homans, *Social Behavior: Its Elementary Forms* (New York: Harcourt, Brace and World, 1961); and Peter M. Blau, *Exchange and Power in Social Life* (New

York: Wiley, 1964) See also G. L. S. Shackle, *Time in Economics* (Amsterdam: North-Holland, 1958).

52. The relationship of sexual development to perceptions of time may be observed in the normal fantasies of young boys who often translate the period of sexual capacity into numbers of presumably allotted orgasms. Not unlike the number of years of life, the number of orgasms is believed to be finite, and hence, with each sexual act, the remaining number is reduced. Thus, psychologically, a boy may feel that the sexual act demands that he give up something. At least it is a reminder that time has clicked forward another notch.

53. Bakan, *op. cit.*

54. On these points, see Sebastian deGrazia, *Of Time, Work and Leisure* (Garden City, N.Y.: Doubleday Anchor, 1964).

55. On this point, see T. B. Roby, Utility and Futurity, *Behavioral Science*, 7 (1962): 194–210.

56. *George Herbert Mead on Social Psychology*, Anselm Strauss (Ed.) (Chicago: University of Chicago Press, 1964) pp. 311–312.

57. A total denial of the relationship between time zones would involve drawing the circles out of linguistic order; for example, the future followed by the past followed by the present. No one in the American sample, in fact, drew the circles this way. Nonetheless, we must keep in mind that this order is not a universal linguistic form. In the Indian sample quite a few circle drawings departed from the order of past followed by present followed by future.

58. On a related point, see O. H. Mowrer and A. D. Ullman, Time as a Determinant in Integrative Learning. *Psychological Review*, 52 (1945): 61–90.

59 This notion is perhaps most relevant to marriages between people of different ethnic groups or social classes.

60. This discussion, while not elaborated in this volume, is surely the basis for future work in the field of time perception. One might wish to explore the meaning of time to married couples and, more specifically, the perceptions of so-called interpersonal time; the effect of faith in marriage on development of a sense of an eternal present; the concept of a shared future influenced surely by the birth of children; the effect of extended families on belief in historical time; the effect of having children on perceptions of the historical future; and of course the whole issue of aging. For a scientific discussion of aging, see B. L. Strehler, *Time, Cells and Aging.* (New York: Academic, 1962).

61. On this point, see Bernard Rosen, The Achievement Syndrome: A Psychocultural Dimension of Social Stratification. *American Sociological Review*, 21 (1956): 203–211.

62. See T. J. Cottle, Nobody's Special When They're Poor. *Yale Review*, in press.

63. A brief paragraph from Parsons' writing reveals the complexity of the so-called dilemma of Object Modalities: "When confronting an object in a situation, the actor faces the dilemma of deciding how to treat it. Is he to treat it in the light of what it is in itself or in the light of what it does or might *flow from its action* [emphasis supplied]? This dilemma can be resolved by giving primacy at the relevant selection points, to the qualities aspect of social objects as a focus of orientation, or by giving primacy to the objects' performance and their outcomes." See T. Parsons, The Dilemma of Object Modalities. In *Towards a General Theory of Action*, T. Parsons and E. A. Shils (Eds.) (New York: Harper & Row, 1962), p. 82.

64. *Ibid.*, p. 83.

65. On this and related points, see Eleanor E. Maccoby, *The Development of Sex Differences* (Stanford: Stanford University Press, 1966).

66. On this point, see Georges Gurvitch, *The Spectrum of Social Time* (Dordrecht, Holland: D. Reidel, 1964).

67. Immigrant families manifesting psychological pathology demonstrate patterns in which women gain positions of control and responsibility, while men frequently reveal an incapacity to adapt to the new culture. On a related point, see S. N. Eisenstadt, *From Generation to Generation* (Glencoe, Ill.: Free Press, 1956).

68. Cited in Arthur Koestler, *The Act of Creation* (New York: Macmillan, 1964), p. 175.

69. On a related point see M. Siffre, *Beyond Time* (London: Chatto and Windus, 1965).

70. On this point, see Ernest Becker, *The Denial of Death* (New York: Free Press, 1974); see also Cumming and Henry, *op. cit.*, and Peter Kastenbaum, *Vitality of Death* (Westport, Connecticut: Greenwood Press, 1971).

71. In this regard Robert Redfield wrote: "One must think that every world view included some spatial and temporal orientation: the cosmos has extension, duration and periodicity." Redfield, *The Primitive World and Its Transformations* (Ithaca, N.Y.: Cornell University Press, 1953), p. 93

72. See Dwayne Huebner, *Curriculum as Concern For Man's Temporality*. Paper read at Curriculum Theory Frontiers, Colloquium Sponsored by Ohio State University, Columbus, Ohio, May 5, 1967.

73. Erikson, *op. cit.*

74. On this point, see Max Heirich, The Use of Time in the Study of Social Change. *American Sociological Review*, **29** (1964): 386–397

BIBLIOGRAPHY

Aaronson, Bernard S., Behavior and the Place Names of Time. *American Journal of Hypnosis.* **9,** No. 1 (July 1, 1966), 1–17.

Achelis, Elisabeth, *Of Time and the Calendar.* New York: Hermitage House, 1955.

Adams, J. A., *Human Memory.* New York: McGraw-Hill, 1967.

Adler, Alfred, *The Neurotic Constitution.* Tr. by Bernard Gluick. New York: Moffatt, Yard, and John E. Lind., London: Kegan Paul, Trench; Trubner and Company Ltd. 1921.

―――― *Problem of Neurosis.* Phillippe Mairet (Ed.). New York: Cosmopolitan, 1930.

―――― *Understanding Human Nature.* Tr. by W. Beran Wolfe. New York: Greenberg, 1924.

Alexander, Franz, *Our Age of Unreason.* Philadelphia: Lippincott, 1942.

―――― The Influence of Psychological Factors upon Gastrointestinal Disturbances. *Psychoanalytic Quarterly,* **3,** (1934), 501–588.

Alexander, Hubert G., *Time as Dimension and History.* Albuquerque: University of New Mexico Press, 1945.

Alexander, Samuel, *Space, Time and Deity.* London, Macmillan, 1920.

Allport, Gordon, *Pattern and Growth in Personality.* New York: Holt, Rinehart & Winston, 1963. (Original copyright, 1937.)

Ames, L., The Development of the Sense of Time in the Young Child. *Journal of Genetic Psychology,* **68** (1946), 97–125.

Anderson, J. C. and Whitely, P. L. The Influence of Two Different Interpolations upon time Estimation. *Journal of Genetic Psychology,* **4** (1930), 391–401.

Angyall, A., *Foundations for a Science of Personality.* New York: Commonwealth Fund, 1941.

Arietti, S., The Processes of Expectation and Anticipation. *Journal of Nervous and Mental Disorders,* **106** (1947), 471–481.

Axel, R., Estimation of Time. *Archives of Psychology,* **12,** No. 74 (1924).

Battle, E. S. and Rotter, J. B., Children's Feelings of Personal Control as Related to Social Class and Ethnic Group. *Journal of Personality,* **31** (1963), 482–490.

Back, K. W. and Gergen, K. J., Apocalyptic and Serial Time Orientations and the Structure of Opinions. *Public Opinion Quarterly,* **27** (1963), 427–442.

Bakan, David, *The Duality of Human Existence.* Chicago: Rand McNally, 1966.

―――― *Of Psychology and Religion* (tentative title, preliminary draft). Lectures in Course at University of Chicago, 1964.

―――― and Kleba, F., Reliability of Time Estimates. *Perceptual and Motor Skills,* **7** (1957), 23–24.

Banfield, Edward, *The Unheavenly City.* Boston: Little, Brown, 1968.

Barndt, R. J. and Johnson, D. N., Time Orientation in Delinquents. *Journal of Abnormal and Social Psychology*, 51 (1955), 343–345.

Barrett, William, The Flow of Time. In Richard M. Gale (Ed.), *The Philosophy of Time*. Garden City, N.Y.: Doubleday Anchor, 1967, pp. 354–376.

Barth, Fredrik, On the Study of Social Change. *American Anthropologist*, 69, No. 6 (December 1967), 661–669.

Bartlett, F. C., *Remembering: A Study in Experimental and Social Psychology*. Cambridge: Cambridge University Press, 1932.

Baruk, H., *La Désorganization de la personnalité*. Paris: Presses Université de France, 1952.

Battle, E. S. and Potter, J. B. Children's Feelings of Personal Control as Related to Social Class and Ethnic Group. *Journal of Personality*, 31 (1963), 482–490.

Becker, Ernest, *The Denial of Death*. New York: Free Press, 1974.

Beckett, Samuel, *Proust*. New York: Grove Evergreen, 1957. (Original publication, 1931.)

Benford, F. "Apparent Time Acceleration with Age, *Science*, 99 (1944), 37.

Berger, G., A Phenomenological Approach to the Problem of Time. In *Readings in Existential Phenomenology*, N. Lawrence and D. O'Connor (Eds.). Englewood Cliffs, N. J.: Prentice-Hall, 1967, pp. 148–204.

Bergler, Edmund and Roheim, Geza, Psychology of Time Perception. *Psychoanalytic Quarterly*, 15 (1946), 190–206.

Bergson, Henri, *Being and Time*. New York: Harper and Row, 1962.

Bergson, Henri, *Duration and Simultaneity, with Reference to Einstein's Theory*. Tr. by Leon Jacobson. Indianapolis: Bobbs-Merrill, 1965.

—— *Matter and Memory*. New York: Humanities, 1970.

—— *Time and Free Will*. London: Allen and Unwin, 1910.

Berman, A., The Relation of Time Estimation to Satiation. *Journal of Experimental Psychology*, 25 (1939), 281–293.

Binswanger, Ludwig, *Being-in-the-World*. Tr. and with an introduction by Jacob Needleman. New York: Basic Books, 1963.

—— The Case of Ellen West. In *Existence*, R. May, E. Angel, and H. F. Ellenberger (Eds.). New York: Basic Books, 1958.

Birren, J. E., The Significance of Age Changes in Speed of Perception and Psychomotor Skills. In *Psychological Aspects of Aging*. Washington: American Psychological Association, 1956.

Blackhan, H. J. *Six Existentialist Thinkers*. New York: Harper & Row, 1952.

Blau, Peter M., *Exchange and Power in Social Life*. New York: Wiley, 1964.

Blauner, Robert, *Alienation and Freedom: The Factory Worker and his Industry*. Chicago: University of Chicago Press, 1964.

Blum, H. F., *Time's Arrow and Evolution*. Princeton: Princeton University Press, 1955.

Boas, George, *The Acceptance of Time*. University of California Publications in Philosophy, 16, No. 12. Berkeley: University of California Press, 1950.

Bock, Philip K., Social Time and Institutional Conflict. *Human Organization*, 25 (Summer 1966), 96–102.

Bonaparte, Marie, Time and the Unconscious. *International Journal of Psychoanalysis*, 21 (1940), 427–468.

Bonier, R. T., *A Study of the Relationship between Time Perspective and Open-Closed Belief Systems*. Unpublished Masters thesis, Michigan State University Library, 1957.

Boss, Medard, *Psychoanalysis and Daseinanalysis*. Tr. by Ludwig B. LeFebre. New York: Basic Books, 1963.

—— *The Analysis of Dreams*. New York: Philosophical Library, 1958.

Bradley, N. C., The Growth of the Knowledge of Time in Children of School Age. *British Journal of Psychology*, 38 (1947), 67–78.

Brandon, S. G. F., *History, Time and Deity*. Manchester: Manchester University Press, 1965.

Brim, Orville, G., Jr. and Wheeler, Stanton, *Socialization After Childhood: Two Essays*. New York: Wiley, 1966.

Brower, J. F. and Brower, D., The Relation between Temporal Judgment and Social Competence in the Feebleminded. *American Journal of Mental Deficiency*, 51 (1947), 619–623.

Brown, Fred, Changes in Sexual Identification and Role Over a Decade and Their Implications. *Journal of Psychology*, 77 (March, 1977), 229–251.

Brumbaugh, Robert, Logic and Time. *Review of Metaphysics*, 18 (June, 1965), 656.

Bruner, Jerome S., *Inhelder and Piaget's The Growth of Logical Thinking: I. A. Psychologist's Viewpoint*. British Journal of Psychology, 50 (1959), 363–370.

——and Goodman, C. C., Value and Need as Organizing Factors in Perception. *Journal of Abnormal and Social Psychology*, 42 (1947), 33–44.

—— and Tajfel, H., Cognitive Risk and Environmental Change. *Journal of Abnormal and Social Psychology*, 62 (1961) 231–241.

Buber, Martin, *Between Man and Man*. Boston: Beacon Press, 1947.

—— *I and Thou*, 2nd ed. Tr. by Ronald G. Smith. New York: Scribner, 1958.

Buck, J. N., The Time Appreciation Test. *Journal of Applied Psychology*, 30 (1946), 388–398.

Buckley, Jerome, *The Triumph of Time*. Cambridge: Harvard University Press, 1966.

Burton, A. A Further Study of the Relation of Time Estimation to Monotony. *Journal of Applied Psychology*, 27 (1943), 350–359.

Burg, J. B., *The Idea of Progress*. New York: Dover, 1955.

Calabresi, R. and Cohen, J., Personality and Time Attitudes. *Journal of Abnormal Psychology*, 73 (1968), 431–439.

Callahan, John F., *Four Views of Time in Ancient Philosophy*. (Cambridge, Harvard University Press, 1948.

Campbell, E. Q., Adolescent Socialization. In *Handbook of Socialization Theory and Research*. David A. Goslin (Ed.). Chicago: Rand McNally, 1969, pp. 821–859.

Campbell, J., Functional Organization of the Central Nervous System with Respect to Orientation in Time. *Neurology*, 4 (1954), 295–300.

Campos, L. P., Relationship between Time Estimation and Retentive Personality Traits. *Journal of Perceptual and Motor Skills*, 23 (1966), 59–62.

Cartwright, L., *To Make A Difference*, New York: Harper & Row, 1967. Quoted in *Time* Magazine, June 9, 1967.

Cassirer, Ernst, *The Philosophy of Symbolic Forms*, New Haven: Yale University Press, 1957.

Chambers, Juanita, Maternal Deprivation and the Concept of Time in Children. *American Journal of Orthopsychiatry*, 31 (1961), 406–416.

Child, I. L., Personality. *Annual Review of Psychology*, 5 (1954), 149–170.

Clay, E. R., *The Alternative: A Study in Psychology*. London, 1882.

Clemence, G. M., Time and Its Measurement. *American Scientist*, 40 (1952), 260–269.

Cohen, John, *Psychological Time in Health and Disease.* Springfield, Ill.: Thomas, 1967.

────── Subjective Time. In *The Voices of Time,* J. T. Fraser (Ed.). New York: Braziller, 1966, pp. 257–275.

────── Experience of Time. *Acta Psychologica,* 10 (1954), 207–219.

────── Disturbances in Time Discrimination in Organic Brain Disease. *Journal of Nervous and Mental Disorders,* 112 (1950), 121–129.

────── Hansel, C. E. M., and Sylvester, J., An Experimental Study of Comparative Judgments of Time. *British Journal of Psychology,* 45 (1954), 108–114.

Collingwood, Robin George, *The Principles of Art.* Oxford: Clarendon, 1955.

Cooper, Joel, Eisenberg, Linda, Robert, John, and Dohrenwend, Barbara S., The Effect of Experimenter Expectancy and Preparatory Effort on Belief in the Probable Occurrence of Future Events. *Journal of Social Psychology,* 71 (1967), 221–226.

Cortes, J. B., The Achievement Motive in the Spanish Economy between the 13th and 18th Centuries. *Economic Development and Cultural Change,* 9 (1960), 144–163.

Coser, L. A. and Coser, R. L., Time Perspective and Social Structure. In *Modern Sociology: An Introduction to the Study of Human Interaction,* A. W. Gouldner and H. P. Gouldner (Eds.). New York: Harcourt, Brace and World, 1963, pp. 638–647.

Cottle, Thomas J., The Felt Sense of Studentry. *Interchange,* 5, 1974, pp. 31–41.

────── Nobody's Special When They're Poor. *Yale Review,* in press.

────── and Howard, Peter, Temporal Extension and Time Zone Bracketing in Indian Adolescents. *Journal of Perceptual and Motor Skills,* 28 (1969), 599–612.

────── and Klineberg, Stephen L., *The Present of Things Future: Explorations of Time in Human Experience.* New York: Free Press, 1974.

────── and Messé, L. A. *Personality Correlates of Attitudes toward Psychotherapy.* Working paper of Counseling Center, Department of Psychology, University of Chicago, 1965.

────── and Muller, John, *An Empirical Investigation of the Intention.* Unpublished manuscript, Department of Social Relations, Harvard University, 1967.

────── Pleck, Joseph, and Kakar, Sudhir, The Time and Content of Significant Life Experiences. *Journal of Perceptual and Motor Skills,* 27 (1968), 155–171.

Cromer, Richard, *The Growth of Temporal Reference during the Acquisition of Language.* Unpublished Ph.D. thesis prospectus, Department of Social Relations, Harvard University, 1967.

Cumming, E. and Henry, W. E., *Growing Old.* New York: Basic Books, 1961.

Curtis, J. H., Duration and Temporal Judgement. *American Journal of Psychology,* 27 (1916), 1–46.

Davids, A., Kidder, C., and Reich, M., Time Orientation in Male and Female Juvenile Delinquents. *Journal of Abnormal and Social Psychology,* 64 (1962), 239–240.

────── and Parenti, Anita N., Time Orientation and Interpersonal Relations of Emotionally Disturbed and Normal Children. *Journal of Abnormal and Social Psychology,* 57 (1958), 299–305.

Davidson, G. M. A., Syndrome of Time Agnosia. *Journal of Nervous and Mental Disorders,* 94 (1941), 336–343.

Decroly, O., and Degand, J., Observations relatives au development de la notion du temps chez petite fille. *Archives de Psychologie* 13 (1913), 113–161.

deGrazia, Sebastian, *Of Time, Work and Leisure.* Garden City, N. Y.: Doubleday Anchor, 1964.

de Tocqueville, Alexis, *Democracy in America*, Vol. 2. New York. Knopf, 1954.

de Unamano, Miguel, *Tragic Sense of Life*. Tr. by J. E. Crawford Flitch. New York: Dover, 1954.

Dewey, John, Time and Individuality. In *Time and Its Mysteries*, H. Shapley (Ed.). New York: Collier, 1962.

—— *Experience and Nature*. New York: Dover, 1958. (Original printing, 1929.)

—— *Human Nature and Conduct*. New York: Holt, 1922.

Dewolfe, R. K. S. and Duncan, C. P., Time Estimation as a Function of Level of Behavior of Successive Tasks. *Journal of Experimental Psychology*, 58 (1959), 153–158.

Dickstein, L. S. and Blatt, S. J., Death Concern, Futurity and Anticipation. *Journal of Consulting Psychology*, 30 (1966), 11–17.

Doob, Leonard W., *Patterning of Time*. New Haven: Yale University Press, 1971.

—— *Becoming More Civilized. A Psychological Exploration*. New Haven: Yale University Press, 1960.

Dooley, Lucille, The Concept of Time in Defense of Ego Integrity. *Psychiatry*, 4 (1941), 13–23.

Dummett, Michael, Bringing About the Past. In *The Philosophy of Time*, R. M. Gale (Ed.). Garden City, N. Y.: Doubleday Anchor, 1967.

Duncan, Otis Dudley, (Editor). *William F. Ogburn on Culture and Social Change*. Chicago: University of Chicago Press, 1964.

Dunkel, Harold B., *Whitehead on Education*, Columbus: Ohio State University Press, 1965.

Dunlap, K., Rhythm and Time. *Psychological Bulletin*, 13 (1916), 206–208.

Dunne, J. W., *This Serial Universe*. London: Macmillan, 1938.

Durkheim, Emile, *The Elementary Forms of Religious Life*. Tr. by Joseph W. Swain. Glencoe, Illinois: The Free Press, 1954.

Eddington, A. S., *Space, Time and Gravitation*. New York: Harper Torchbook, 1959.

Efron, Robert, The Duration of the Present. In *Interdisciplinary Perspectives of Time*, Roland Fischer (Consulting Ed.), *Annals of the New York Academy of Sciences*, 138 (1967), 713–729.

Einstein, Albert, *Relativity*, New York: Crown, 1961.

Eisenstadt, S. N., *From Generation to Generation*. Glencoe, Ill.: Free Press, 1956.

Eissler, K. R., Time and the Mechanism of Isolation. *Psychoanalytic Review*, 39 (1952), 1–22.

Eliade, Mircea, *The Sacred and the Profane*. New York: Harper& Row, 1961.

—— *Cosmos and History: The Myth of the Eternal Return*. New York: Harper & Row, 1959.

—— Time and Eternity in Indian Thought. In *Man and Time*, J. Campbell (Ed.). New York: Pantheon, 1957.

Elkine, D., De l'Orientation de l'enfant d'age scolaire dans les relations temporelles. *Journal de Psychologie Normale et Pathologique*, 28 (1928), 425–429.

Epley, D. and Ricks, D. F., Foresight and Hindsight in the T.A.T. *Journal of Projective Techniques*, 27, No. 1 (1963), 51–59.

Erikson, Erik H., *Identity, Youth and Crisis*. New York: Norton, 1968.

—— *Insight and Responsibility*. New York: Norton, 1964.

——Identity and the Life Cycle. *Psychological Issues*, 1, No. 1 (1959), Whole.

—— *Young Man Luther*. New York: Norton, 1958.

—— *Childhood and Society.* New York: Norton, 1950.

Eson, Morris E., *An Analysis of Time Perspectives at Five Age Levels.* Unpublished Ph. D. dissertation. University of Chicago, 1951.

Eson, Morris E. and Greenfield, Norman, Life Space: Its Content and Temporal Dimension. *Journal of Genetic Psychology,* **100** (1962), 113–128.

—— and Kafka, J. S., Diagnostic Implications of a Study in Time Perception. *Journal of General Psychology,* **46** (1952), 169–183.

Falk, J. L. and Bindra, D., Judgement of Time as a Function of Serial Position and Stress. *Journal of Experimental Psychology,* **47** (1954), 274–284.

Farber, M. L., Suffering and Time Perspective of the Prisoner. In *Authority and Frustration.* University of Iowa Studies of Child Welfare, **20** (1944), 153–227.

Farber, M. L., Time Perspective and Feeling Tone: A Study in the Perception of the Days. *Journal of Psychology,* **35** (1953), 253–257.

Farnham-Diggory, Sylvia, Self, Future and Time: A Developmental Study of the Concepts of Psychotic, Brain-Damaged and Normal Children. *Monographs of the Society for Research in Child Development,* **31**, No. 1 (1966). Whole.

Fennichel, Otto, *Psychoanalytic Theory of Neurosis.* New York: W. W. Norton, 1945.

Filer, R. J. and Meals, D. W., The Effect of Motivating Conditions on the Estimation of time. *Journal of Experimental Psychology,* **39** (1949), 327–337.

Fingarette, Herbert, *The Self in Transformation.* New York: Basic Books, 1963.

Fink, H. H., The Relationship of Time Perspective to Age, Institutionalization and Activity. *Journal of Gerontology,* **12** (1957), 414–417.

Firey, Walter, Conditions for the Realization of Values Remote in Time. In *Sociocultural Theory, Values and Sociocultural Change,* Edward A Tiryakian (Ed.). Glencoe, Ill.: Free Press, 1963.

Fisher, S. and Fisher, R. L., Unconscious Conception of Parental Figures as a Factor Influencing Perception of Time. *Journal of Personality,* **21** (1953), 496–505.

Fordhan, Frieda, *An Introduction to Jung's Psychology.* London: Penguin, 1953.

Fraisse, Paul, *The Psychology of Time.* New York: Harper & Row, 1963.

—— Of Time and the Worker. *Harvard Business Review,* **137** (May–June 1959), 121–125.

——Étude comparée la perception et de l'estimation de la durée chez les enfants et chez les adultes. *Enfance,* **1** (1948), 199–211.

—— and Oleron, G., La Perception de la durée d'un son d'intensité croissante. *Année Psychologie,* **50** (1950), 327–343.

Franck, Kate and Risen, E. A., A Projective Test of Masculinity-Femininity. *Journal of Consulting Psychology,* **13** (1949), 247–256.

Frank, Lawrence K., Time Perspectives. *Journal of Social Philosophy,* **4** (1939), 293–312.

Frankl, Viktor E., Time and Responsibility. *Existential Psychiatry* No. 3 (Fall 1966), 361–366.

Fraser, J. T., *Of Time, Passion and Knowledge.* New York: Braziller, 1975.

—— (Ed.), *Voices of Time.* New York: Braziller, 1965.

Freud, Sigmund. A Disturbance of Memory on the Acropolis and Screen Memories. In *Collected Papers,* Vol. V. London: Hogarth, 1950.

——*Group Psychology and the Analysis of the Ego.* Tr. by James Strachey, New York: Bantam, 1936.

────── *The Problem of Anxiety.* New York: Norton, 1936.

────── *New Introductory Lectures on Psychoanalysis.* New York: Norton, 1933.

Friedman, K. C., A Time Comprehension Test. *Journal of Educational Research,* **39** (1945, 62–68.

────── Time Concepts of Elementary School Children. *Elementary School Journal,* **44** (1943–1944, 337–352.

Fried, Marc, *The World of the Urban Working Class.* Cambridge: Harvard University Press, 1973.

Friedman, Maurice, *To Deny Our Nothingness.* New York: Delacorte, 1967.

Fromm, Erich, *Escape from Freedom.* New York: Rinehart, 1941.

Fox, G., The Effects of Internal and Duration of Visual Stimuli upon the Time Error. *Psychological Newsletter,* **45** (1952), 1–17.

General Classification Test, Adjutant General's Office. *Psychological Bulletin,* **42** (1945), 760–768.

Gendlin, E. T. and Shlien, J. M., Immediacy in Time Attitudes before and after Time-Limited Psychotherapy. *Journal of Clinical Psychology,* **17** (1961), 69–72.

Gesell, A. and Ilg, F. L., *Infant and Child in the Culture of Today.* New York: Harper, 1943.

Getzels, Jacob and Jackson, Donald, *Creativity and Intelligence.* New York: Wiley, 1962.

Gilliland, A. R., Hofeld, J. B., and Eckstrand, G., Studies in Time Perception. *Psychologial Bulletin,* **43** (1946), 162–176.

Gioscia, Victor, Adolescence, Addiction, and Achrony. In *Personality and Social Life,* Robert Endleman (Ed.). New York: Random House, 1967.

────── *Classification of Family Types.* Unpublished manuscript, Department of Psychology, Queens College, 1966.

Glenn, Evelyn, *Volition: A Study of Factors Related to Level of Experienced Volition.* Unpublished. Harvard University, Thesis prospectus, Department of Social Relations, 1967.

Gordon, C. and K. J. Gergen (Eds.), *The Self in Social Interaction.* New York, Wiley, 1968.

Gorman, B. S. and Katz, B., Temporal Orientation and Anality. *Proceedings of the 79th Annual Convention, American Psychological Association,* 1971, 367–368. Washington, D.C.

Gorman, B. S., Wessman, A. E., Schmeidler, G. R., Thayer, S., and Mannucci, E., Linear Representation of Temporal Location and Stevens' Law. *Memory and Cognition,* **1** (1973), 169–171.

Gothberg, L. C. The Mentally Defective Child's Understanding of Time. *American Journal of Mental Deficiencies,* **53** (1948), 441–455.

Gough, H. G., Identifying Psychological Femininity. *Educational and Psychological Measurement,* **12** (1952), 427–439.

Graves, T. D. *Time Perspective and the Deferred Gratification Pattern in a Tri-Ethnic Community.* Unpublished Ph.D. dissertation, University of Pennsylvania, 1962.

Grebe, J. J., Time: Its Breadth and Depth in Biological Rhythms. *Annals of New York Academy of Science,* **112** (1962), 1206–1210.

Grinker, Roy, Treatment of War Neurosis. *Journal of the American Medical Association,* **126,** No. 3 (1944), 142–145.

────── and Spiegel, J. P., *Men under Stress,* Philadelphia: Blackiston, 1945.

Gilford, J. P., Spatial Symbols in the Apprehension of Time. *American Journal of Psychology*, **37** (1926), 420–423.

Gulliksen, H., The Influence of Occupation upon the Perception of Time. *Journal of Experimental Psychology*, **10** (1927), 52–59.

Gurvitch, Georges, *The Spectrum of Social Time*. Dordrecht, Holland: D. Reidel, 1964.

——— Social Structure and the Multiplicity of Times. In *Sociological Theory, Values and Sociocultural Change*, Edward A. Tiryakian (Ed.). Glencoe, Ill. Free Press, 1963, pp. 171–184.

Haberman, P. W., The Use of Psychological Tests for Recall of Past Situations. *Journal of Clinical Psychology*, **19** (1963), 245–249.

Haggerty, Lee J., Another Look at the Burgess Hypothesis: Time as an Important Variable. *American Journal of Sociology* **76** (May 1971), 1084–1093.

Hall, Edward T., *The Silent Language*. Garden City, N.Y.: Doubleday, 1959.

Hammerschmidt, William W., *Whitehead's Philosophy of Time*. New York: King's Crown, 1947.

Hare, R. D., Psychopathy, Fear Arousal and Anticipated Pain. *Psychological Reports*, **16** (1965), 499–502.

Hartmann, Heinz, *Essays on Ego Psychology*, New York: International Universities Press, 1964.

Harton, John J., The Influence of the Difficulty of Activity on the Estimation of Time. *Journal of General Psychology*, **21** (1939), 51–62.

Harton, J. J., An Investigation of the Influence of Success and Failure on the Estimation of Time. *Journal of Genetic Psychology*, **21** (1939), 51–62.

Hatcher, Anne, *Universals in John Updike*. Unpublished honors thesis, Department of Social Relations, Harvard University, 1967.

Hearnshaw, L. W., Temporal Integration and Behavior. *Bulletin of the British Psychological Society*, **30** (1956), 1–20.

Heidegger, Martin, *Being and Time*. New York: Harper & Row, 1962.

Heirich, Max, The Use of Time in the Study of Social Change. *American Sociological Review*, **29** (1964), 386–397.

Henry, William, *The Analysis of Fantasy*. New York: Wiley, 1956.

——— The Business Executive: The Psychodynamics of a Social Role. *American Journal of Sociology*, **54** (1949), 286–291.

Hering, Daniel W., The Time Concept and Time Sense Among Cultured and Uncultured Peoples. In *Time and Its Mysteries*. H. Shapley (Ed.). New York: Collier, 1962.

Hilgard, E. R. Human Motives and the Concept of the Self. *American Psychologist*, 4 (1949), 374–382.

Hindle, H. M., Time Estimates as a Function of Distance Travelled and Relative Clarity of a Goal. *Journal of Personality*, **19** (1951), 483–490.

Hoagland, H., The Chemistry of Time. *Science Monitor*, **56** (1943), 56–61.

——— The Physiological Control of Judgements of Duration: Evidence for a Chemical Clock. *Journal of Genetic Psychology*, **9** (1933), 267–287.

Hoffer, A. and Osmond, H. The Relationship between Mood and Time Perception. *Psychiatric Quarterly Supplement*, **36**, No. 1 (1962), 87–92.

Homans, George C., *Social Behavior: Its Elementary Forms*. New York: Harcourt Brace and World, 1961.

Horney, Karen, *Our Inner Conflicts, A Constructive Theory of Neurosis.* New York: Norton, 1945.

—— *The Neurotic Personality of Our Time.* New York: Norton, 1937.

Huebner, Dwayne, *Curriculum as Concern for Man's Temporality.* Unpublished manuscript, 1967.

Hume, David, *A Treatise on Human Nature,* 2 Vols. T. H. Green and T. H. Grose (Eds.). London: Longmans, Green, 1874.

Hunter, W. S., Delayed Reactions in Animals and Children. *Behavior Monographs,* **2,** No. 1 (1913).

Iggers, George, G., The Idea of Progress in Recent Philosophies of History. *Journal of Modern History,* **30,** No. 3 (1958), 215–226.

Inhelder, B., and Piaget, J., *The Growth of Logical Thinking From Childhood to Adolescence.* Tr. by A. Parsons and S. Milgram. New York: Basic Books, 1958.

Inkeles, Alex, Social Structure and Socialization. In *Handbook of Socialization Theory and Research.* David A. Goslin (Ed.). Chicago: Rand McNally, 1969.

—— Society, Social Structure, and Child Socialization. In *Socialization and Society.* John Clausen (Ed.). Boston: Little, Brown, 1968.

Israeli, N., *Abnormal Personality and Time.* New York: Science, 1936.

Jahoda, M., Some Socio-Psychological Problems of Factory Life. *British Journal of Psychology,* **31** (1941), 191–206.

James, William, *Psychology: The Briefer Course* (1892). New York: Harper Torchbooks, 1961.

—— *Essays in Radical Empiricism.* London: Longmans Green, 1912.

—— *Principles of Psychology,* 2 Vols. London: Macmillan, 1891.

Janet, Pierre, *L'Évolution de la memoire et la notion de temps.* Paris: Chahine, 1928.

Jung, Carl G., *Modern Man in Search of a Soul.* New York: Harcourt Brace, 1933.

Kagan, Jerome, Acquisition and Significance of Sex Typing and Sex Role Identity. *Review of Child Development Research.* New York: Russell Sage Foundation, 1964, 137–167.

——The Child's Sex Role Classification of School Objects. In Martin L. Hoffmann and Lois W. Hoffman (Eds.). *Child Development,* **35** (1964), 1051–1056.

—— The Child's Perception of the Parent. *Journal of Abnormal and Social Psychology,* **53** (1956), 257–258.

—— and Moss, Howard, *From Birth to Maturity.* New York: Wiley, 1962.

Kahn, P., Time Orientation and Reading Achievement. *Journal of Perceptual and Motor Skills,* **21** (1965), 157–158.

Kant, Immanual, *Critique of Pure Reason.* New York: Dutton, 1934.

Kaplan, Joel R., Ludwig Binswanger's Existential Analysis. *Existential Psychiatry,* **6,** No. 2 (Summer 1967), 239–256.

Kastenbaum, Peter, *Vitality of Death.* Westport, Connecticut: Greenwood, 1971.

Kastenbaum, Robert, Cognitive and Personal Futurity in Later Life. *Journal of Individual Psychology,* **19** (1963), 216–222.

—— The Dimensions of Future Time Perspective, an Experimental Analysis. *Journal of General Psychology,* **65** (1961), 201–218.

—— Time and Death in Adolescence. In *The Meaning of Death,* H. Feifel (Ed.). New York: McGraw-Hill, 1959, Chapter 7.

Katz, Daniel, The Functional Approach to the Study of Attitudes. *Public Opinion Quarterly*, 24 (1960), 163–204.

——— and Stotland, E. *Functional Theory of Attitudes.* In *Psychology: A Study of a Science*, S. Koch (Ed.), Vol III. New York: McGraw-Hill, 1959.

Kelly, George, *The Psychology of Personal Constructs.* 2 Vols. New York: Norton, 1955.

Ketchum, J. D., Time, Values and Social Organization. *Canadian Journal of Psychology*, 5 (1951), 97–109.

Kierkegaard, Soren, *The Concept of Dread.* Tr. by Walter Lowrie. Princeton: Princeton University Press, 1944. (Originally published in Danish, 1844.)

——— *Sickness Unto Death.* Tr. by Walter Lowrie. Princeton: Princeton University Press, 1941. (Originally published in Danish, 1844.)

Kimberly, R. P., Rhythmic Patterns in Human Interaction. *Nature*, 228 (1970), 88–90.

King, Arden R., Time, Society and Culture. *Human Mosaic*, 3 (1968), 1–11.

King, Magda, *Heidegger's Philosophy.* New York: Dell, 1964.

Klineberg, Otto, *Social Psychology*, 2nd ed. New York: Holt, 1954.

Klineberg, Stephen L., Changes in Outlook on the Future Between Childhood and Adolescence. *Journal of Personality and Social Psychology*, 7, No. 2 (1967), 185–193.

——— *The Outlook on the Future in Childhood and Adolescence.* Unpublished thesis prospectus, Harvard University, 1965.

——— *Structure of the Psychological Future.* Unpublished thesis, Harvard University, 1966.

Kluckhohn, Florence R., Dominant and Variant Value Orientations. In F. R. Kluckhohn, and Fred L. Strodtbeck (Eds.). *Variations in Value Orientations.* Evanston, Ill.: Row, Peterson, 1961.

——— and Strodtbeck, F., (Eds.) *Variations in Value Orientations.* Evanston, Ill.: Row, Peterson, 1961.

Knapp, Robert H. Attitudes Towards Time and Aesthetic Choice. *Journal of Social Psychology*, 1960.

——— and Garbutt, John T., Time Imagery and the Achievement Motive. *Journal of Personality*, 26 (1958), 423–434.

Knapp, Robert H. and Garbutt, John T., Variation in Time Descriptions and in Achievement. *Journal of Social Psychology*, 67 (1965), 269–272.

——— and Green, Helen B., The Judgment of Music-Filled Intervals and Achievement. *Journal of Social Psychology*, 54 (1961), 263–269.

Koestler, Arthur, *The Act of Creation.* New York: Macmillan, 1964.

Kohler, W., *Gestalt Psychology.* New York: Liveright, 1947.

Kowalski, W. J., The Effect of Delay upon the Duplication of Short Temporal Intervals. *Journal of Experimental Psychology*, 33 (1943).

Koyré, Alexandre, *From the Closed World to the Infinite Universe.* Baltimore: Johns Hopkins, 1957.

Kracauer, Siegfried, Time and History. In *History and the Concept of Time*, Middletown, Conn.: Wesleyan University Press, 1966, pp. 65–78.

Krauss, H. H., Anxiety: The Dread of a Future Event. *Journal of Individual Psychology*, 23, (1967), 88–93.

——— and Ruiz, R. A., Anxiety and Temporal Perspective. *Journal of Clinical Psychology*, 23 (1967), 454–455.

Kubler, George, *The Shape of Time.* New Haven: Yale University Press, 1962.

Kuhlen, R. G. and Johnson, G. H., Changes in Goals with Increasing Adult Age. *Journal of Consulting Psychology,* 16 (1952), 1–4.

Kummel, Friedrich, Time as Succession and the Problem of Duration. In *The Voices of Time,* J. T. Fraser (Ed.). New York: Braziller, 1966.

Langer, J., Wapner, S., and Werner, H., The Effect of Danger upon the Experience of Time, *American Journal of Psychology,* 74 (1961), 94–97.

Langer, S., *Philosophy in a New Key.* Cambridge: Harvard University Press, 1957.

Lansky, L. M., Mechanisms of Defense: Sex Identity and Defenses Against Aggression. In *Inner Conflict and Defense,* D. R. Miller and G. E. Swanson (Eds.). New York: Holt, 1960, pp. 272–288.

Lavelle, L., *Du Temps et de l'éternité.* Paris: Aubier, 1945.

Lavin, David, *The Prediction of Academic Performance.* New York: Russell Sage Foundation, 1965.

Leary, Timothy, *Interpersonal Diagnosis of Personality: A Functional Theory and Methodology for Personality Evaluation.* New York: Ronald, 1957.

Lecomte Du Nouy, P., Apparent Time Acceleration with Age. *Science,* 99 (1944), 38.

——— *Human Destiny.* New York: David McKay, 1947.

Leshan, Lawrence, Time Orientation and Social Class. *Journal of Abnormal Psychology,* 47 (1957), 589–592.

Lessing, Elise, *Developmental Personality and Other Correlates of Length of Future Time Perspective (FTP).* Unpublished research report, Institute for Juvenile Research, 4 No. 2 (1967).

——— Demographic, Developmental, and Personality Correlates of Length of Future Time Perspective. *Journal of Personality,* 36, No. 2, (June 1967), 183–201.

Levine, Donald, *Wax and Gold: Tradition and Innovation in Ethopian Culture.* Chicago: University of Chicago Press, 1965.

Levine, M., Spivack, G., Fushchillo, J., and Tavernier, A. Intelligence and Measures of Inhibition and Time Sense. *Journal of Clinical Psychology,* 15 (1959), 224–226.

Levi, A. W., *Philosophy and the Modern World.* Bloomington: Indiana University Press, 1959.

Lewin, Kurt *A Dynamic Theory of Personality.* New York: McGraw-Hill, 1935.

Lewin, Kurt, *Resolving Social Conflicts: Selected Papers on Group Dynamics.* New York: Harper and Bros., 1948.

Lewin, Kurt. Time Perspective and Morale. In Watson, *Civilian Morale.* Boston: Houghton Mifflin, 1942, pp. 48–70.

Lewis, A., The Experience of Time in Mental Disorder. *Proceedings of the Royal Society of Medicine,* 25, No. 1 (1931–1932), 611–620.

Lewis, Wyndham, *Time and Western Man.* Boston: Beacon, 1957.

Lhamon, W. T. and Goldstone, S., The Time Sense. *American Medical Association Archives of Neurological Psychiatry,* 76 (1956), 625–629.

Lifton, Robert, Youth and History: Individual Change in Postwar Japan. *Daedalus,* Winter, 91 1962, 172–197.

Lindzey, G., Review of Lewin's Field Theory in Social Science. *Journal of Abnormal and Social Psychology,* 47 (1952), 132–133.

Lipman, R. S., *Some Relationships between Manifest Anxiety, Defensiveness, and Future Time Perspective*. Paper read at Eastern Psychological Association, 1957. Cited by M. Wallace and A. I. Rabin, Temporal Experience. *Journal of Abnormal and Social Psychology*, 57 (1960), 213–236.

Lipsitt, Paul D., *Defensiveness in Decision Making as a Function of Sex Role Identification*. Unpublished manuscript, 1966.

Lovejoy, Arthur O., *The Reason, the Understanding, and Time*. Baltimore: John Hopkins, 1961.

Luscher, Kurt K., Time: A Much Neglected Dimension in Social Theory and Research. *Sociological Analysis and Theory*, 4 (October 1974): 101–116.

Lynn, D. B., A Note on Sex Differences in the Development of Masculine and Feminine Identification. *Psychological Review*, 66 (1959), 126–135.

Maccoby, Eleanor E., *The Development of Sex Differences*. Stanford: Stanford University Press, 1966.

MacIver, Robert M., *The Challenge of the Passing Years: My Encounter with Time*. New York: Trident, 1962.

MacKinnon, D. W., A Topological Analysis of Anxiety. *Character and Personality*, 12 (1944), 163–176.

Mahrer, A. P., The Role of Expectancy in Delayed Reinforcement. *Journal of Experimental Psychology*, 52 (1965), 101–105.

Malamud, Bernard, Man in the Drawer. *Atlantic*, 221, No. 4 (April 1968).

Malrieu, P., *Les Origines de la conscience du temps*. Paris: Presses Université de France, 1953.

Maltz, Daniel N., Primitive Time-Reckoning as a Symbolic System. *Cornell Journal of Social Relations*, 3 (1968), 85–113.

Maltzman, I., Fox, J., and Morrisett, L., Some Effects of Manifest Anxiety on Mental Set. *Journal of Experimental Psychology*, 46 (1953), 50–54.

Mann, Thomas, Voyage with Don Quixote. In *Essays by Thomas Mann*. Tr. by E. T. Lowe-Porter. New York: Knopf, 1957, pp. 325–369.

Mannheim, Karl, *Essays on Sociology and Social Psychology*. New York: Oxford University Press, 1953.

Maslow, Abraham, *Toward a Psychology of Being*. Princeton, N.J.: Van Nostrand, 1962.

May, Rollo. Contributions of Existential Psychotherapy. In *Existence: A New Dimension in Psychiatry and Psychology*, Rollo May, Ernst Angel, and Henri Ellenberger (Eds.). New York: Basic Books, 1958.

———— *The Meaning of Anxiety*. New York: Ronald, 1950.

———— Angel, Ernst, and Ellenberger, Henri (Eds.), *Existence: A New Dimension in Psychiatry and Psychology*. New York: Basic Books, 1958.

McArthur, Charles, Personality Differences between Middle and Upper Classes. *Journal of Abnormal and Social Psychology*, 2 (1955), 247–255.

McClelland, David C., *The Achieving Society*. Princeton, N.J.: Van Nostrand, 1961.

———— *Personality*. New York: Holt, Rinehart & Winston, 1951.

———— et. al., *Talent and Society*. Princeton, N.J.: Van Nostrand, 1958.

McDougall, R., Sex Differences in the Sense of Time. *Science*, 19 (1904), 707–708.

McDougall, W., *Outline of Psychology*. New York: Scribner, 1923.

McLuhan, Marshall. *Space, Time and Poetry*. Explorations 4 (1955), 56–62.

McTaggart, J. M. E., Time. In *The Philosophy of Time*, R. M. Gale (Ed.). Garden City, N.Y.: Doubleday Anchor, 1967.

Mead, George Herbert, *Mind, Self and Society*, Chicago: University of Chicago Press, 1934.

Mead, Margaret. *New Lives for Old*. New York: Morrow, 1956.

Meade, Robert D., Time Estimates as Affected by Motivational Level, Goal Distance and Rate of Progress. *Journal of Experimental Psychology*, 58 (October 1959), 275–279.

Meals, D. and Filer, R., The Effect of Motivating Conditions on the Estimation of Time. *Journal of Experimental Psychology*, 39 (1949), 327–331.

Meerloo, Joost A. M., The Time Sense in Psychiatry. In *The Voices of Time*, J. T. Fraser (Ed.). New York: Braziller, 1966.

———— *The Two Faces of Man: Two Studies on the Sense of Time and on Ambivalence*. New York: International Universities Press, 1954.

Melges, Frederick Towne and Fourgerousse, Carl Edward, Time Sense, Emotions, and Acute Mental Illness. *Journal of Psychiatric Research*, 4, No. 2 (November 1966), 127–139.

Merlan, Philip, Brentano and Freud. *Journal of the History of Ideas*, 6 (1945), 375–77, and 10 (1949), 451.

———— Time Consciousness in Husserl and Heidegger. *Philosophy and Phenomenological Research*, 8, No. 1 (September 1947).

Merleau-Ponty, Maurice, *Phenomenology of Perception*. New York: Humanities, 1962.

Meyerhoff, Hans, *Time in Literature*. Berkeley: University of California Press, 1955.

Michaud, E., *Essai sur l'organization de la connaissance entre 10 et 14 ans*. Paris: Urin, 1949.

Miller, D. R. and Swanson, G. E., *Inner Conflict and Defense*. New York: Holt, 1960.

Minkowski, E., Bergson's Conceptions as Applied to Psychopathology. *Journal of Nervous and Mental Disorders*, 33 (1926), 553–568.

———— *Le Temps vecu*, L'evolution psychiatrique, Paris: Centre d'Editions Psychiatriques, 1933.

Minkowski, E., Étude psychologique et analyse phenomenologique d'un cas de melancolie schizophrenique. *Journal de Psychologie Normale et Pathologique*, 20 (1923), 543–558. Tr. by Barbare Bliss as: Findings in a Case of Schizophrenic Depression. In *Existence: A New Dimension in Psychiatry and Psychology*, R. May, E. Angel and H. F. Ellenberger (Eds.). New York: Basic Books, 1958, pp. 127–138.

Mischel, W., Delay of Gratification, Need for Achievement, and Acquiescence in Another Culture. *Journal of Abnormal and Social Psychology*, 62 (1961), 543–552.

———— Preference for Delayed Reinforcement and Social Responsibility. *Journal of Abnormal and Social Psychology*, 62 (1961), 1–7.

———— Preference for Delayed Reinforcement: An Experimental Study of a Cultural Observation. *Journal of Abnormal and Social Psychology*, 56 (1958), 57–61.

Moore, Wilbert E., *Man, Time and Society*. New York: Wiley, 1963.

———— *Social Change*. Englewood Cliffs, N.J.: Prentice-Hall, 1965.

Morgenstern, I. *The Dimensional Structure of Time*. New York: Philosophical Library, 1950.

Mowrer, O. H. and Ullman, A. D. Time as a Determinant in Integrative Learning. *Psychological Review*, 52 (1945), 61–90.

Muller, John, *Self, Time and Activity*. Unpublished manuscript, Department of Social Relations, Harvard University, 1967.

——— *The Temporal Structuring of Competence*. Unpublished manuscript, Department of Social Relations, Harvard University, 1966.

Mundt, Ernest, *Art, Form and Civilization*. Berkeley: University of California, 1952.

Munitz, Milton K., *Space, Time and Creation: Philosophical Aspects of Scientific Cosmology*. Glencoe, Ill.: Free Press, 1957.

Murphy, G., *Personality: A Biosocial Approach to Origins and Structure*. New York: Harper, 1947.

Murray, Henry A., Preparations for the Scaffold of Comprehensive System. In *Psychology: A Study of Science: Formulations of the Person and the Social Context*, Sigmund Koch (Ed.). Vol. III, New York: McGraw-Hill, 1959, pp. 5–54.

——— *Explorations in Personality: A Clinical and Experimental Study of Fifty Men of College Age*. New York: Oxford University Press, 1938.

Mussen, Paul and Rutherford, Eldrid, Parent–Child Relations and Parental Personality in Relation to Young Children's Sex Role Preference. *Child Development*, **34**, No. 3 (1963), 589–607.

Nakajima, S., On the Time Error in the Successive Comparison of Time. *Japanese Journal of Psychology*, **21** (1951), 36–45.

Niebuhr, Richard N., *The Responsible Self*. New York: Harper & Row, 1963.

Oackden, E. C. and Sturt, M., The Development of the Knowledge of Time in Children. *British Journal of Psychology*, **12** (1922), 309–337.

Oberndorf, C. P., Time, Its Relation to Reality and Purpose. *Psychoanalytic Review*, **7** (1941), 139–155.

Ogburn, W. F., Stationary and Changing Societies. *American Journal of Sociology*. **42** (July 1936), 16–32.

Oleron, G., Influence de l'intensité d'un son sur l'estimation de sa durée apparente. *Année Psychologie*, **52** (1952), 383–392.

Orme, J. E., *Time, Experience and Behavior*. New York: American Elsevier, 1969.

Ornstein, R. E., *On the Experience of Time*. Baltimore: Penguin, 1970.

Osgood, Charles, The Nature and Measurement of Meaning. *Psychological Bulletin*, **49** (1952), 197–237.

——— and Suci, G. J., Factor Analysis of Meaning. *Journal of Experimental Psychology*, **50** (1955), 325–338.

——— Suci, G. J., and Tannenbaum, P. H., *The Measurement of Meaning*. Urbana: University of Illinois Press, 1957.

Park, R. E., *Human Communities: The City and Human Ecology*. Glencoe, Ill.: Free Press, 1952.

Parkman, Margaret, *Identity, Role and Family Functioning*. Unpublished Ph.D. disseration, Department of Sociology, University of Chicago, 1965.

Parsons, Talcott, Youth in the Contest of American Society. *Daedalus*, **91** (Winter 1962), pps. 97–123.

Parsons, Talcott, *Youth in the Context of American Society Daedalus*, **91** (Winter 1962), pps. 97–123.

———and Bales, R. F., *Family, Socialization and Interaction*. New York: Free Press, 1955.

——— Bales, R. F. and Shils, E. A., *Working Papers in the Theory of Action*. Glencoe, Ill.: Free Press, 1953.

Pascal, Blaise, *Pensées of Pascal*. New York: Peter Pauper, 1946.

Persky, H., Grinker, R. R., Mirsky A., and Gamm, S. R., *Life Situations, Emotions and the Excretions of Hippuric Acid in Anxiety States*. Research publication, Association of Nervous and Mental Disease, **29** (1950), 297–306.

Pettit, T. Anality and Time. *Journal of Consulting and Clinical Psychology*, **33** (1969), 170–174.

Piaget, Jean, Time Perception in Children. In *The Voices of Time*, J. T. Fraser (Ed.). New York: Braziller, 1966.

——— *The Construction of Reality in the Child*. New York: Basic Books, 1954.

——— The Development of Time Concepts in the Child. In *Psychopathology of Childhood*, P. H. Hoch and J. Zubin (Eds.). New York: Grune and Stratton, 1955.

——— *Épistémologie génétique*, 3 Vols. Paris: Presses Université de France, 1950.

——— Les Notions de mouvement et la vitesse chez l'enfant. Paris: Presses Université de France, 1946.

Pistor, F., Measuring the Time Concepts of Children. *Journal of Educational Research*, **33** (1939), 293–300.

Platt, J. J., Eisenman, R., and DeGross, E., Birth Order and Sex Differences in Future Time Perspective. *Developmental Psychology*, **1** (1969), 70.

Plutchik, R., *The Emotions: Facts, Theories and a New Model*. New York: Random House, 1962.

Polak, Fred L. *The Image of the Future*, 2 vols. New York: Oceana, 1961.

Poulet, G., *Studies in Human Time*. Tr. by E. Coleman. New York: Harper & Row, 1956.

Priestly, J. B., *Man and Time*. Garden City, N.Y.: Doubleday, 1964.

Rapaport, David, States of Consciousness: A Psychopathological and Psychodynamic View (1951). In *The Collected Papers of David Rapaport*, Merton M. Gill (Ed.). New York: Basic Books, 1967, pp. 385–404.

Redfield, Robert, *The Primitive World and Its Transformations*. Ithaca, N.Y.: Cornell University Press, 1953.

Reichenbach, H. (Ed.), *The Direction of Time*. Berkeley: University of California Press, 1956.

Richmond, B., *Time Measurements*. Leiden: Brill, 1956.

Ricks, D. F., and Umbarger, C., and Mack, R. A Measure of Increased Temporal Perspective in Successfully Treated Adolescent Delinquent Boys. *Journal of Abnormal and Social Psychology*, **69** (1964), pp. 685–689.

Riesman, David, *The Lonely Crowd*. New Haven: Yale University Press, 1950.

Roberts, Alan, and Hermann, Robert S., Dogmatism, Time Perspective and Anomie. *Journal of Individual Psychology*, **16** (1960), 67–72.

Roby, T. B., Utility and Futurity. *Behavioral Science*, **7** (1962), 194–210.

Roelofs, O. and Zeeman, W. P. G., The Subjective Duration of Time Intervals. *Acta Psychologica*, **6** (1549), 126–177.

Rogers, Carl, *Client Centered Therapy*. Boston: Houghton Mifflin, 1951.

——— The Place of the Person in the New World of the Behavioral Sciences. *Personal Guidance Journal*, **39** (1961), 442–451.

—*A Theory of Therapy, Personality, and Interpersonal Relationships as Developed in the Client Centered Framework.* Counseling Center Library, University of Chicago, 1956.

Rokeach, Milton, *The Open and Closed Mind.* New York: Basic Books, 1960.

Rosen, B. The Achievement Syndrome: A Psychocultural Dimension of Social Stratification. *American Sociological Review,* 21 (1956), 203–211.

———— and D'Andrade, R. G., The Psychosocial Origins of Achievement Motivation. *Sociometry,* 22 (1959), 185–218.

Rosenzweig, S. and Koht, A. G., The Experience of Duration as Affected by Need Tension. *Journal of Experimental Psychology,* 16 (1933), 745–774.

Ross, S. and Katchmar, L., The Construction of a Magnitude Function for Short-Time Intervals. *American Journal of Psychology,* 37 (1951), 211–214.

Roth, J. A. *Timetables.* Indianapolis: Bobbs-Merrill, 1963.

Ruiz, R. A., Reivich, R. S. and Krauss, H. H., Tests of Temporal Perspective: Do They Measure the Same Construct? *Psychological Reports,* 21 (1967), 849–852.

Russell, Bertrand. *The Problems of Philosophy.* New York: Oxford University Press, 1959.

———— *History of Western Philosophy.* New York: Simon and Schuster, 1945.

———— On the Experience of Time, *Monist,* 25 (1915), 212–233.

Rychlak, J. E. Manifest Anxiety as Reflecting Commitment to the Psychological Present at the Expense of Cognitive Futurity. *Journal of Consulting and Clinical Psychology,* 38 (1972), 70–79.

Saint Augustine, *Confessions.* Tr. by J. G. Pilkington. New York: Dial, 1964.

Schachtel, Ernest G., *Metamorphosis.* New York: Basic Books, 1959.

———— On Memory and Childhood Amnesia. *Psychiatry,* 10 (1947), 1–26.

Schilder, P., Psychopathology of Time. *Journal of Nervous and Mental Disorders,* 83 (1936), 530–546.

Schultz, John, Custom. In *4 x 4.* New York: Grove, 1962.

Schutz, Alfred, *Collected Papers: The Problem of Social Reality,* Vol. I. The Hague: Martinus Nijhoff, 1962.

Sechrest, Lee, The Psychology of Personal Constructs. In *Concepts of Personality,* Joseph M. Wepman and Ralph W. Heine (Eds.). Chicago: Aldine, 1963.

Shackle, G. L. S. *Time in Economics.* Amsterdam: North-Holland, 1958.

Sherover, Charles M. *Heidegger, Kant and Time.* Bloomington: Indiana University Press, 1971.

Shils, Edward A. Primordial, Personal, Sacred and Civil Ties. *British Journal of Sociology,* 81 (1957), 130–146.

Siegman, A. W., The Relationship Between Future Time Perspective, Time Orientation, and Impulse Control in a Group of Young Offenders and in a Control Group. *Journal of Consulting Psychology,* 25 (1961), 470–475.

Siffre, M., *Beyond Time.* London: Chatto and Windus, 1965.

Singer, J. L., Delayed Gratification and Ego Development: Implications for Clinical and Experimental Research. *Journal of Consulting Psychology,* 19 (1955), 259–266.

Skalet, M., The Significance of Delayed Reations in Young Children. *Comparative Psychology Monographs,* 7, No. 4 (1930–1931).

Smart, J. J. C., *Problems of Space and Time.* New York: Macmillan, 1964.

Smith, John E., *The Spirit of American Philosophy*. New York: Oxford University Press, 1966.

Smith, Marian W., Different Cultural Concepts of Past, Present and Future. *Psychiatry*, **15** (1952), 395–400.

Smith, Robert J., Cultural Differences and the Concept of Time. In *Aging and Leisure*, Robert W. Kleemeier (Ed.). New York: Oxford University Press, 1961.

Smythe, Elizabeth and Goldstone, S., The Time Sense: A Normative Genetic Study of the Development of Time Perception. *Journal of Perceptual and Motor Skills*, **7** (1957), 49–59.

Sollberger, Arne, General Properties of Biological Rhythms. *Annals of the New York Academy of Sciences*, **98**, No. 4 (1962), 757–774.

Sorokin, Pitriam A., *Sociocultural Causality, Space, Time*. Durham, N. C.: Duke University Press, 1943.

———— and Merton, R. K., Social Time: A Methodological and Functional Analysis, *American Journal of Sociology*, **42** (March 1937), 615–629.

Spengler, Oswald, *The Decline of the West*. New York: Knopf, 1926.

Spreen, O., The Position of Time Estimation in a Factor Analysis and its Relationship to some Personality Variables. *Psychological Record*, **13**, No. 4 (1963), 455–464.

Starr, Chester G., Historical and Philosophical Time. In *History and the Concept of Time*. Middletown, Conn.: Wesleyan University Press, 1966, pp. 24–35.

Stein, K. B., Sarbin, T. R., and Kulik, J. A., Future Time Perspective: Its Relation to the Socialization Process and the Delinquent Role. *Journal of Consulting and Clinical Psychology*, **32** (1968), 257–264.

Stekel, William, *Conditions of Nervous Anxiety and Their Treatment*. Tr. by Rosalie Gabler. London: Kegan, Paul, Trench, Trubner, 1923.

Straus, Ervin W., Existential Approach to Time. In *Interdisciplinary Perspectives of Time*, Roland Fischer (Consulting Ed.). *Annals of the New York Academy of Sciences*, **138** (1967), 759–766.

Strauss, Anselm (Ed.). *George Herbert Mead on Social Psychology*. Chicago: University of Chicago Press, 1964.

Strauss, E. S., Disorders of Personal Time in Depressive States. *Southern Medical Journal*, **40**, No. 3, 154–158.

Strehler, B. L., *Time, Cells and Aging*. New York: Academic, 1962.

Strodtbeck, Fred L., Family Interaction, Values and Achievement. In D. C. McClelland et al. (Eds.). *Talent and Society*. Princeton, N.J.: Van Nostrand, 1958.

———— Bezdek, William and Goldhammer, Donald, *Personal Efficacy, Problem Severity and Willingness to Act to Reduce Water Pollution*. Unpublished manuscript, Social Psychology Laboratory, University of Chicago, 1966.

Sturt, Mary, *The Psychology of Time*. New York: Harcourt and Brace, 1925.

———— Experiments on the Estimate of Duration. *British Journal of Psychology*, **13** (1923), 382–388.

Sullivan, Harry Stack, The Theory of Anxiety and the Nature of Psychotherapy. *Psychiatry*, **2**, No. 1 (1949), 3–13.

———— The Meaning of Anxiety in Psychiatry and Life. *Psychiatry*, **2**, No. 1 (1948), 1–15.

Swanson, G. E., Determinants of the Individual's Defenses Against Inner Conflict. In

Parental Attitudes and Child Behavior, J. C. Glidewell (Ed.). Springfield, Ill. Thomas, 1961, pp. 5–41.

Swift, E. Y. and McGeoch, J. A., An Experimental Study of the Perception of Filled and Empty Time. *Journal of Experimental Psychology*, 8 (1925), 240–249.

Syngg, D. and Combs, A., *Individual Behavior: A New Frame of Reference for Psychology*. New York: Harper, 1948.

Taylor, Janet, Theory and Manifest Anxiety. *Psychological Bulletin*, 53 (1969), 303–320.

——— A Personality Scale of Manifext Anxiety. *Journal of Abnormal and Social Psychology*, 48 (1953), 285–290.

Teahan, J. E., Future Time Perspective, Optimism and Academic Achievement. *Journal of Abnormal and Social Psychology*, 57 (1958), 379–380.

Temperly, Nicholas, M., Personal Tempo and Subjective Accentuation. *Journal of General Psychology*, 68, No. 2 (1963), 267–287.

Thayer, S., Gorman, B. S., Wessman, A. E., Schmeidler, G., and Mannucci, E., The Relationship Between Locus of Control and Temporal Experience. *Journal of Genetic Psychology*, in press.

Thor, D. H., Diurnal Variability in Time Estimation. *Journal of Perceptual and Motor Skills*, 15, No. 2 (1962), 451–454.

Tillich, Paul, *Theology of Culture*. New York: Oxford University Press, 1964.

Tiryakian, Edward A. The Existential Self and the Person. In C. Gordon and K. J. Gergen (Eds). *The Self in Social Interaction*. New York: Wiley, 1968.

——— (Ed.), *Sociological Theory, Values and Sociocultural Change*. Glencoe, Ill.: Free Press, 1963.

——— *Sociologism and Existentialism*. Englewood Cliffs, N.J.: Prentice-Hall, 1962.

Tompkins, Sylvan S., *Affect-Imagary-Consciousness*, 2 Vols. New York: Springer, 1962.

Toulmin, Stephen and Goodfield, June, *The Discovery of Time*. New York: Harper Torchbooks, 1966.

Toynbee, Arnold J., *Change and Habit: The Challenge of Our Time*. New York: Oxford University Press, 1966.

Tymieniecka, Anna-Teresa, *Phenomenology and Science in Contemporary European Thought*. New York: Farrar, Straus and Cudahy, 1962.

Updike, John, *Pigeon Feathers*. Greenwich, Conn.: Fawcett, 1963.

van der Leeuw, G., Primordial Time and Final Time. In *Man and Time*, J. Campbell (Ed.). New York: Pantheon, 1957.

Vincent, J. W., *An Investigation of Constricted and Extended Temporal Perspectives*. Unpublished Master's thesis, University of Oregon, 1965.

——— and Tyler, L. E. A Study of Adolescent Time Perspectives. *Proceedings of the 73rd Annual Convention of the American Psychological Association*, Chicago, Illinois, 1965.

Voegelin, Eric, Immortality: Experience and Symbol. *Harvard Theological Review*, 60 (1967).

Von Bertalarffy, Ludwig, *Robots, Men and Minds; Psychology in the Modern World*. New York: Braziller, 1967.

Wallace, Melvin, Future Time Perspective in Schizophrenia. *Journal of Abnormal and Social Psychology*, 52 (1956), 240–245.

——— and Rabin, Albert, Temporal Experience, *Psychological Bulletin*, 57 (1960), 213–236.

Wallis, Robert, *Time: Fourth Dimension of the Mind*. Tr. by Betty B. Montgomery and Denis B. Montgomery. New York: Harcourt, Brace and World, 1966.

Warm, Joel S., Morris, James R., and Kew, John K., Temporal Judgment as a Function of Nosological Classification and Experimental Method. *Journal of Psychology*, **55**, No. 2 (1963), 287–297.

Warner, W. Lloyd, *Yankee City*, one vol., abridged ed. New Haven: Yale University Press, 1963.

───── and Henry, W. Radio Day Time Serial: A Symbolic Analysis. *Genetic Psychology Monograph*, **37**, No. 10 (1948), 3–71.

Weber, Alden O., Estimation of Time. *Psychological Bulletin*, **30** (1933), 233–252.

Weber, Max, *The Protestant Ethic and the Spirit of Capitalism*. Tr. by Talcott Parsons. New York: Scribner, 1958.

Wechsler, David, *The Measurement and Appraisal of Adult Intelligence*. Baltimore: Williams and Wilkins, 1958.

Weinstein, Fred and Platt, Gerald M., *The Wish To Be Free*. Berkley: University of California Press, 1969.

Werner, Heinz, *Comparative Psychology of Mental Development*, rev. ed. Chicago: Follett, 1948.

Wessman, Alden E. Personality and the Subjective Experience of Time. *Journal of Personality Assessment*, **37**, No. 2 (1973), 103–114.

───── and Ricks, David F., *Mood and Personality*. New York: Holt Rinehart & Winston, 1966.

White, H. L. and Edmundson, M. S. Notes on Coupling and Decoupling. Unpublished manuscript, Harvard University, Department of Social Relations, 1967.

White, Robert W., *Lives in Progress*. New York: Holt, Rinehart and Winston, 1966.

───── Ego and Reality in Psychoanalytic Theory: A Proposal Recording Independent Ego Energies. *Psychological Issues*, **3**, No. 3. Mono. II (1963).

───── Motivation Reconsidered: The Concept of Competence. *Psychological Review*, **66** (1959), 297–333.

Whitehead, Alfred North, *Adventures of Ideas*. New York: Free Press, 1967.

───── *Symbolism: Its Meaning and Effect*. New York: Putnam, Capricorn Books, 1959.

───── *Modes of Thought*. Glencoe: The Free Press, 1968.

───── *Process and Reality*. Glencoe, Illinois: Free Press, 1969.

Whitrow, G. J., *What is Time*. London: Thames and Hudson, 1972.

───── *The Natural Philosophy of Time*. New York: Harper & Row, 1961.

Winterbottom, Marian, P., The Relation of Need for Achievement to Learning Experiences in Independence and Mastery. In J. W. Atkinson (Ed.), *Motives in Fantasy, Action, and Society*. Princeton, N.J.: Van Nostrand, 1958, 453–478.

Winthrop, Henry, The Sociologist and the Study of the Future. *The American Sociologist*, **3** (May 1968), 136–145.

Witkin, H. A., *Psychological Differentiation: Studies of Development*. New York: Wiley, 1962.

Wittgenstein, Ludwig, *The Blue and Brown Books*. New York: Harper Torchbook, 1958.

Wohlford, Paul F., Extension of Personal Time, Affective States, and Expectation of Personal Death, *Journal of Personality and Social Psychology*, **3** (1966), 559–566.

—— *Determinants of Extension of Personal Time.* Unpublished Ph.D. dissertation, Duke University, 1964.

Wyndham, Lewis, *Time and Western Man.* Boston: Beacon, 1957.

Yaryan, R. B. and Festinger, L., Preparatory Action and Belief in the Probable Occurrence of Future Events. *Journal of Abnormal and Social Psychology,* **63** (1961), 603–606.

Yates, Sybille, Some Aspects of Time Difficulties and Their Relation to Music. *International Journal of Psycho-Analysis,* **16** (1938), 341–354.

Yerkes, R. M., and Urban, Time Estimation in its Relation to Sex, Age, and Physiological Rhythms. *Harvard Psychological Studies,* **2** (1906), 405–430.

Young, M., and Ziman, J., Cycles in Social Behavior. *Nature* **229** (1971), 91–95.

Zeigarnik, B. V., *The Pathology of Thinking.* Tr. by Basil Haigh. New York: Consultant Bureau, 1965.

Zeller, A. F., Experimental Analogue of Repression. *Journal of Experimental Psychology,* **40** (1950), 411–423.

Zentner, H. The Social-Time-Space Relationship: A Theoretical Formulation. *Sociological Inquiry,* **36** (1966), 61–79.

Zimbardo, Philip G., Marshall, Gary, and Maslach, Christina, Liberating Behavior from Time Bound Control: Expanding the Present Through Hypnosis. *Journal of Applied Social Psychology,* **1,** No. 4 (1971), 305–323.

TABLES

Table 1 Percentage Distributions of Experiences in the Five Time Zones

Number of Experiences	Distant Past		Near Past		Present		Near Future		Distant Future	
	Males	Females	Males	Females	Males	Females	Males	Females	Males	Females
0	9.2	12.9	4.7	6.9	41.1	23.6	41.8	33.3	54.5	57.8
1	21.7	17.8	8.5	14.7	30.7	33.3	28.5	35.4	21.0	21.6
2	19.1	25.8	13.9	13.7	13.2	20.7	14.9	13.7	10.9	11.8
3	20.1	17.8	20.1	14.7	7.8	11.8	8.4	12.7	8.3	6.9
4	11.8	6.9	17.5	18.6	3.3	5.9	4.0	2.9	4.0	0.0
5	9.0	8.9	14.4	14.7	1.4	2.9	1.6	2.0	0.95	2.0
6	5.0	5.9	11.1	11.8	1.9	0.98	0.71	0.0	0.24	0.0
7	1.6	2.0	5.3	4.9	0.48	0.98	0.24	0.0	0.24	0.0
8	1.4	2.0	2.6	0.0	0.0	0.0	0.0	0.0	0.0	0.0
9	0.95	0.0	1.9	0.0	0.0	0.0	0.0	0.0	0.0	0.0
10	0.0	0.0	0.0	0.0	0.0	0.0	0.0	0.0	0.0	0.0
Total	423	101	423	102	423	102	423	102	423	102
Mean	2.7[a]	2.6[a]	3.8	3.4	1.2	1.6	1.1	1.2	0.91	0.75
Standard deviation	1.92[a]	1.93[a]	2.05	1.95	1.41	1.46	1.35	1.23	1.28	1.11

[a] Number of experiences per zone.

Table 2 Intercorrelation Matrix of all Six Experiential Inventory Variables for Men (N = 423) and woman (N = 102)

Experiential Variables	Experiential Mean		Number of Distant-Past Experiences		Number of Near-Past Experiences		Number of Present Experiences		Number of Near-Future Experiences		Number of Distant-Future Experiences	
	M	F	M	F	M	F	M	F	M	F	M	F
Experiential mean	···	···										
Number of distant-past experiences	-.64**	-.60**	···	···								
Number of near-past experiences	-.32**	-.19**	-.26**	-.30**	···	···						
Number of present experiences	.24**	.24*	-.29**	-.23*	-.24**	-.29**	···	···				
Number of near-future experiences	.63**	.64**	-.35**	-.38**	-.34**	-.27**	-.06	.03	···	···		
Number of distant-future experiences	.71**	.65**	-.23**	-.20*	-.40**	-.26**	-.13	-.16	.18*	.18	···	···

*p < .05, one-tailed test.
**p < .01, one-tailed test.

217

Table 3 Frequency Distributions of Duration Responses for Men and Women

Duration variables	1 Seconds		2 Minutes		3 Hours		4 Days		5 Weeks		6 Months		7 Years	
	%	N	%	N	%	N	%	N	%	N	%	N	%	N
Males														
Present backward	47.5	143	9.3	28	7.0	21	8.0	24	6.0	18	7.6	23	13.9	42
Present forward	42.3	127	10.7	32	7.7	23	6.0	18	4.3	13	9.7	29	18.3	55
Distant past	0.67	2	2.0	6	1.0	3	1.0	3	0.67	2	5.0	15	89.3	268
Near past	7.7	23	3.0	9	2.7	8	6.3	19	10.0	30	62.3	187	7.7	23
Distant future	2.0	6	2.0	6	1.0	3	2.0	6	2.7	8	6.6	20	83.4	251
Near future	12.6	38	4.0	12	3.0	9	8.0	24	13.6	41	48.2	145	10.3	31
Females														
Present backward	40.4	38	5.3	5	7.4	7	7.4	7	9.6	9	8.5	8	21.3	20
Present forward	29.0	27	8.6	8	11.8	11	3.2	3	7.5	7	15.0	14	24.7	23
Distant past	1.0	1	0.0	0	0.0	0	0.0	0	4.2	4	7.4	7	87.4	83
Near past	2.1	2	3.2	3	3.2	3	6.3	2	9.5	9	66.3	63	9.5	9
Distant future	2.1	2	1.0	1	0.0	0	1.0	1	4.2	4	11.6	11	80.0	76
Near future	7.4	7	3.2	3	5.3	5	8.4	8	15.8	15	51.6	49	8.4	8

Table 4   Percentages and Totals of Symmetrical and Asymmetrical Perceptions of the Present

Present in Advance of Itself (Negative Values)[a]				Symmetrical Present (Zero Values)				Present Behind Itself (Positive Values)			
Males		Females		Males		Females		Males		Females	
%	N	%	N	%	N	%	N	%	N	%	N
26.5	89	34.5	32	59.8	201	56.9	53	13.7	46	8.6	8

[a] By nature of the calculations, negative values indicate that the present's future extension is lower than its past extension.

Table 5   Distribution of Experiential Mean Scores Divided by Past, Present, and Future Orientations for Men ($N = 423$) and Women ($N = 102$)

Experiential Inventory Variables	Past Orientation		Present Orientation		Future Orientation	
	Males	Females	Males	Females	Males	Females
Mean-score range	1.0–1.9		2.0–2.4		2.5–5.0	
%	30.3	22.5	51.3	60.8	18.4	16.7
Total	128	23	217	62	78	17

Table 6 Relationship of Experiential Orientations and Duration Responses of the Present

Duration Variable	Males						Females					
	Past[a]	Present[a]	Future[a]	%	χ^2	df	Past[a]	Present[a]	Future[a]	%	χ^2	df
Present forward												
Instantaneous (seconds–minutes)	51 (51)	91 (53)	30 (48)	51.7			8 (42)	23 (40)	2 (12)	37.6		
Middle (hours–weeks)	22 (22)	30 (18)	13 (21)	19.3	0.869	4	3 (16)	17 (29)	3 (19)	22.6	8.53**	4
Extended (months–years)	27 (27)	50 (29)	19 (31)	29.0			9 (42)	18 (31)	11 (69)	39.8		
Present backward												
Instantaneous (seconds–minutes)	59 (59)	94 (55)	34 (54)	56.2			9 (45)	30 (52)	4 (25)	45.7		
Middle (hours–weeks)	20 (20)	38 (22)	14 (23)	21.6	0.48	4	4 (20)	15 (26)	4 (25)	24.5	5.61**	4
Extended (months–years)	21 (21)	39 (23)	14 (23)	22.2			7 (35)	13 (22)	8 (50)	29.8		
Total	100	171	62				20	58	16			
%	30.0	51.4	18.6				21.3	61.7	17.0			

[a] The number of respondents in a particular category. Percentages are given in parentheses.

*$p > .05$.

**$p > .10$.

Table 7 Relationship of Experiential Orientations and Duration Responses of the Near and Distant Future[a]

| | Experiential Orientation | | | | | | | | | | | |
| | Males | | | | | | Females | | | | | |
Duration Variable	Past[a]	Present[a]	Future[a]	%	χ^2	df	Past[a]	Present[a]	Future[a]	%	χ^2	df
Near future												
Seconds–days	26 (26)	43 (25)	11 (18)	29.3			6 (30)	16 (27)	1 (6)	24.2		
Weeks–months	62 (62)	114 (66)	46 (74)	61.4	8.92*	4	13 (65)	39 (66)	12 (75)	67.4	5.28	4
Years	12 (12)	15 (9)	5 (8)	9.3			1 (5)	4 (7)	3 (19)	8.4		
Distant future												
Seconds	7 (7)	27 (16)	0 (0)	10.5			0 (0)	7 (12)	1 (6)	8.4		
Months	14 (14)	7 (4)	2 (3)	6.9	25.55**	4	2 (10)	8 (14)	1 (6)	11.6	3.82	4
Years	79 (79)	138 (80)	60 (97)	82.6			18 (90)	44 (74)	14 (88)	80.0		
Total	100	172	62				20	59	16			
%	29.9	51.5	18.6				21.1	62.1	16.8			

[a] In making these calculations the criteria delineating divisions in time zones required adjustment because the patterns of response for these time zones did not yield the same bimodal quality found in responses to the present's duration (Table 6).
[b] The number of respondents in a particular category. Percentages are given in parentheses.
*p < .10.
**p < .001.

Table 8 Distribution of Payments for Personal Past

Money Value	Time Allotment											
	Hour				Day				Year			
	Males		Females		Males		Females		Males		Females	
	%	N	%	N	%	N	%	N	%	N	%	N
$ 0	44.4	189	49.5	49	33.8	144	43.4	43	32.7	140	54.0	54
$ 10	16.2	69	13.1	13	11.0	47	11.1	11	5.4	23	5.0	5
$ 100	13.6	58	13.1	13	15.5	66	16.2	16	9.1	39	9.0	9
$ 1,000	7.3	31	9.1	9	16.9	72	12.1	12	12.8	55	10.0	10
$10,000	18.5	79	15.2	15	22.8	97	17.2	17	40.0	171	22.0	22
%	100		100		100		100		100		100	
Total		426		99		426		99		428		100

Table 9 Distribution of Payments for Personal Future

Money Value	Time Allotment											
	Hour				Day				Year			
	Males		Females		Males		Females		Males		Females	
	%	N	%	N	%	N	%	N	%	N	%	N
$ 0	46.6	198	56.0	56	42.4	180	46.0	46	41.5	177	47.5	48
$ 10	15.3	65	17.0	17	6.6	28	11.0	11	2.6	11	2.9	3
$ 100	14.1	60	10.0	10	14.9	63	23.0	23	5.6	24	8.9	9
$ 1,000	5.6	24	8.0	8	13.2	56	9.0	9	6.8	29	11.9	12
$10,000	17.9	76	8.0	8	22.4	95	10.0	10	43.2	184	28.8	29
%	100		100		100		100		100		100	
Total		425		100		424		100		426		101

Table 10 Distribution of Payments for Historical Past

Money Value						Time Allotment						
	Hour				Day				Year			
	Males		Females		Males		Females		Males		Females	
	%	N	%	N	%	N	%	N	%	N	%	N
$ 0	73.7	314	81.0	81	70.7	301	78.0	78	70.2	299	72.3	73
$ 10	7.0	30	6.0	6	6.8	29	5.0	5	4.2	18	4.9	5
$ 100	5.4	23	5.0	5	7.0	30	8.0	8	6.1	26	0.99	1
$ 1,000	2.4	10	2.0	2	4.9	21	6.0	6	3.9	17	5.9	6
$10,000	11.5	49	6.0	6	10.6	45	3.0	3	15.6	66	15.8	16
%	100		100		100		100		100		100	
Total		426		100		426		100		426		101

Table 11 Distribution of Payments for Historical Future

Money Value						Time Allotment						
	Hour				Day				Year			
	Males		Females		Males		Females		Males		Females	
	%	N	%	N	%	N	%	N	%	N	%	N
$ 0	55.8	237	67.5	68	54.1	230	64.0	64	55.0	235	71.0	71
$ 10	5.6	24	4.9	5	1.66	7	4.0	4	4.4	19	0.0	0
$ 100	7.5	32	5.9	6	8.3	35	13.0	13	3.4	14	6.0	6
$ 1,000	6.6	28	5.9	6	11.1	47	8.0	8	3.5	15	6.0	6
$10,000	24.5	104	15.8	16	24.9	106	11.0	11	33.7	144	17.0	17
%	100		100		100		100		100		100	
Total		425		101		425		100		427		100

Table 12   Means and Standard Deviations for the Money Game

Epoch	Allotment	Males		Females	
		Mean	SD	Mean	SD
Personal past	Hour	2.43	1.55	2.29	1.51
	Day	2.88	1.58	2.47	1.55
	Year	3.31	1.73	2.38	1.67
Personal future	Hour	2.37	1.56	1.93	1.32
	Day	2.72	1.67	2.23	1.40
	Year	3.14	1.88	2.70	1.77
Historical past	Hour	1.73	1.37	1.46	1.09
	Day	1.80	1.39	1.51	1.07
	Year	1.92	1.53	1.89	1.55
Historical future	Hour	2.43	1.72	1.99	1.55
	Day	2.57	1.76	1.99	1.45
	Year	2.62	1.87	1.99	1.60

Table 13   The Semantic Differential Instrument

	Very	Somewhat more	Slightly		Slightly	Somewhat more	Very		
hard	:	:	:	:	:	:		soft	(potency)
dirty	:	:	:	:	:	:		clean	(evaluation)
small	:	:	:	:	:	:		large	(potency)
good	:	:	:	:	:	:		bad	(evaluation)
strong	:	:	:	:	:	:		weak	(potency)
cruel	:	:	:	:	:	:		kind	(evaluation)
cowardly	:	:	:	:	:	:		brave	(potency)
fair	:	:	:	:	:	:		unfair	(evaluation)
rugged	:	:	:	:	:	:		delicate	(potency)
unpleasant	:	:	:	:	:	:		pleasant	(evaluation)

Table 14   Means and Standard Deviations of Potency and Evaluation Factors of Time, Past, Present, and Future

	Males				Females			
	Potency		Evaluation		Potency		Evaluation	
Variable	Mean	SD	Mean	SD	Mean	SD	Mean	SD
Time	4.9	0.984	4.4	1.20	4.8	1.07	4.8	1.41
Past	4.5	0.989	4.5	1.52	4.7	1.07	4.2	1.78
Present	4.9	1.04	4.6	1.52	5.0	1.08	5.0	1.56
Future	5.1	1.03	5.2	1.36	5.1	1.15	5.5	1.55

Table 15 Intercorrelations of Semantic Differential Mean Scores of Time, Past, Present, and Future [a]

Variables	Time Potency	Time Evaluation	Past Potency	Past Evaluation	Present Potency	Present Evaluation	Future Potency	Future Evaluation
Time potency	· · ·	.28*	.29**	.14	.44**	.28*	.59**	.33**
Time evaluation	.19**	· · ·	.33**	.38**	.42**	.44**	.39**	.49**
Past potency	.13*	.03	· · ·	.19	.48**	.05	.45**	.21
Past evaluation	−.05	.16*	.00	· · ·	.20	.26**	.22*	.30**
Present potency	.19**	.10	.20**	.18**	· · ·	.17	.64**	.35**
Present evaluation	.03	.41**	.08	.10	.02	· · ·	.26*	.56**
Future potency	.22**	.14*	.09	.11	.38**	.31**	· · ·	.48**
Future evaluation	.02	.26**	.06	.18**	.20**	.38**	.30**	· · ·

[a] Males are below the diagonal, females above.

*p < .05, two-tailed test.

**p < .01, two-tailed test.

Table 16 Frequency Distribution of Responses to Future Commitment Scale

	Agree (%)		Disagree (%)		Can't Say (%)	
Items	Males	Females	Males	Females	Males	Females
1. I will have a good and successful life.	42.6	35.7	.78	1.95	56.6	62.3
2. Russia and America will live in peaceful coexistence in the future.	19.4	11.7	25.8	17.5	54.8	70.8
3. The weather in America will become continually worse in the coming years.	10.6	11.0	26.9	22.1	62.5	66.9
4. Eventually the earth will collide with another planet.	4.4	4.5	30.2	31.7	65.4	63.6
5. Someday the Republican Party will dominate the American political system.	7.2	4.5	34.4	31.8	58.4	63.6
6. Man will someday find life on many other planets.	63.3	55.8	4.6	5.8	32.0	38.3
7. Someday the earth will be so crowded with people that there will not be enough food.	33.1	22.7	34.6	37.7	32.3	39.6
8. Someday man will only have to work one day a week.	16.3	14.9	45.5	42.9	38.2	42.2
9. There will always be racial problems in America.	47.3	47.4	32.8	24.0	19.9	28.6
10. Someday the entire world will be run by the United Nations.	16.0	13.6	46.2	37.7	37.7	48.7
11. Someday cancer will be completely cured.	83.5	71.4	2.6	4.5	13.9	24.0
12. In the future, if a man dies at the age of 100, he will be considered young.	32.8	37.0	26.1	20.8	41.1	42.2

Table 16 Continued

	Agree (%)		Disagree (%)		Can't Say (%)	
Items	Males	Females	Males	Females	Males	Females
13. In the future more and more people will get divorces.	45.7	47.1	19.9	16.9	34.4	35.9
14. In the future there will be so many automobiles that man will have to find a new method of transportation.	57.7	57.8	17.7	16.2	24.7	26.0
15. In my lifetime I will suffer no serious illnesses.	6.7	9.7	14.7	17.5	78.5	72.7
16. In the future the average American family will be getting larger and larger.	20.9	26.6	45.5	37.7	33.6	35.7
17. In the future religion will become less and less important to Americans.	31.0	21.4	50.6	53.9	18.3	24.7
18. I will live a very long life.	13.9	15.6	6.2	6.5	79.8	77.9
19. America will be at war again.	51.7	33.1	7.5	7.8	40.0	59.1
20. There is a limit to how fast airplanes can be made to fly.	33.6	26.0	32.9	33.1	29.5	40.9
21. Cities will always have slums.	61.5	58.4	19.6	21.4	18.9	20.1

228

22.	Someday I will be rich.	11.9	7.1	9.6	16.9	78.5	76.0
23.	In the future many lucky things will happen to me.	23.3	24.7	4.6	5.8	72.1	69.5
24.	The problem of juvenile delinquency will never be solved.	36.9	26.6	38.7	42.9	24.3	30.5
25.	There will always be a war somewhere on the face of the earth.	55.3	59.1	17.0	12.3	27.6	28.6
26.	My children will become greatly successful.	17.7	8.4	.52	1.3	86.8	90.3
27.	I will see all of my children graduate from college.	19.6	14.9	4.9	5.2	75.4	79.9
28.	My spouse will probably outlive me.	29.2	9.1	3.1	5.8	67.7	85.1
29.	Someday the whole world will speak one common language.	24.3	20.1	31.5	28.6	44.2	51.3
30.	My grandchildren will travel to Mars as people travel to Europe.	22.5	14.3	12.7	13.6	64.9	72.1
31.	There is life on the planet Jupiter.	5.9	6.5	18.3	16.2	75.7	77.3

Table 17 Factor Analyses of Future Commitment Scale Items

Item	Imediate Social I	Distant Cosmic Personal (Negativism) II	Mediate Global (Optimism) III	IV	h^2
24. Juvenile delinquency	.69	−.10	.07	−.21	.53
25. War on the earth	.67	−.10	.06	−.13	.48
9. Racial problems	.66	−.12	.03	−.12	.46
21. City slums	.66	−.02	.11	−.16	.48
17. Religious unimportance	.44	.01	.03	−.32	.30
19. America to be at war	.42	−.13	.28	−.22	.32
13. More divorces	.39	−.05	.21	−.31	.29
22. Personal wealth	.16	−.66	.08	.06	.47
26. Successful children	−.10	−.66	.09	−.16	.48
27. Children to be college graduates	−.03	−.62	.13	−.21	.45
1. Personally successful	.05	−.60	−.11	−.18	.40
23. Personally lucky	.27	−.60	.16	−.01	.46
18. Personal long life	.06	−.60	.24	−.04	.42
15. Children's health	.09	−.48	−.01	−.19	.27
3. Worsening U.S. weather	.06	−.01	.69	−.09	.49
4. Earth's collision	−.06	−.11	.55	−.10	.33
2. Peace between U.S.A. and U.S.S.R.	.01	−.06	.52	−.30	.37
5. Republican Party domination	.14	−.10	.51	−.01	.29
7. Earth's crowdedness	.32	.06	.47	−.20	.37
28. Outlived by spouse	.18	−.23	.38	−.14	.25
20. No limit to airplane speed	.25	−.10	.35	−.18	.23
31. Life on Jupiter	.21	−.24	.30	.06	.20
29. World with one language	.11	−.05	.06	−.66	.45
11. Cancer cured	.19	−.13	.09	−.57	.38

Table 17   Continued

Item	Imediate Social I	Personal II	Distant Cosmic (Negativism) III	Mediate Global (Optimism) IV	h^2
10. U.N. to run world	.12	−.09	.17	−.55	.35
12. Extended longevity	.14	−.15	.20	−.53	.36
30. Interplanetary travel	.10	−.19	.27	−.50	.37
16. U.S. family increasing	.20	−.07	−.04	−.50	.30
8. One-day work week	.21	−.14	.14	−.48	.31
14. Too many automobiles	.24	−.06	.24	−.40	.28
6. Life on other planets	.26	−.20	.35	−.40	.38
Sum of squares over variables	3.01	2.93	2.57	3.15	

Table 18   Means and Standard Errors for Future Commitment Scale and Individual Commitment Scales

Future Commitment Scale Variables	Males		Females		Difference	Value
	Mean	Error	Mean	Error		
Immediate social	1.0	.095	1.3	.133	0.3	2.00*
Personal commitment	4.3	.123	4.6	.150	0.3	1.56
Distant cosmic	2.8	.113	3.1	.145	0.3	1.68*
Mediate global	2.2	.128	2.7	.158	0.5	2.43*
Total "can't-say" score	10.4	.332	11.8	.432	1.4	2.59**

*$p < .05$, two-tailed test.
**$p < .01$, two-tailed test.

231

Table 19 Summary of Relationships between Experiential Categories and Personal Prediction, Future Purchasing, Time Potency Rating, and Present Evaluation Rating

Characteristics of Related Temporal Variables	Experiential Orientation									
	Males					Females				
	Future-Avoidant [a]	Middle [a]	Future-Oriented [a]	χ^2	p (2 df)	Future-Avoidant [a]	Middle [a]	Future-Oriented [a]	χ^2	p (2 df)
1. High on personal prediction	63 (54)	52 (42)	48 (35)	4.91	> .05	12 (50)	18 (44)	14 (40)	.391	n.s.
2. Purchasing future ("fantasizers")	58 (50)	62 (50)	57 (42)	2.40	n.s.	7 (28)	24 (59)	16 (48)	5.88	> .05
3. High on time potency	53 (45)	55 (44)	82 (60)	7.56	< .05	10 (40)	24 (59)	18 (52)	1.78	n.s.
4. High on present evaluation	68 (58)	69 (56)	58 (43)	6.05	< .05	14 (55)	23 (56)	16 (48)	0.387	n.s.
Total	117	124	136			25	41	34		
%	31.0	32.9	36.1			25.0	41.0	34.0		

[a] The number of respondents in a particular category. Percentages are given in parentheses.

Table 20 Intercorrelations between the Future Commitment Scale and Duration Items

Duration Items	Future Commitment Scale (31 Items)	
	Males	Females
Past boundary of present	−.08	−.05
Future boundary of present	−.13	.03
Distant past	−.05	−.24*
Near past	−.04	−.07
Distant future	−.06	−.26*
Near future	−.01	−.28*

*$p < .05$, two-tailed.

Table 21 Frequency Distribution of Temporal Relatedness Scores

Temporal Relatedness	Males		Females	
	%	Number	%	Number
Atomistic	60	256	65	66
Continuous	27	115	26	27
Integrated-projected	13	57	9	9
Total	100	428	100	102

Table 22　Frequency Distribution of Dominance Scores

Dominance	Past Percentage	Past Number	Present Percentage	Present Number	Future Percentage	Future Number
Males						
Absence (0 points)	65	276	39	168	20	87
Secondary (2 points)	27	116	52	222	14	59
Dominance (4 points)	8	36	9	38	66	282
Total	100	428	100	428	100	428
Females						
Absence (0 points)	72	73	23	24	26	27
Secondary (2 points)	22	22	67	68	9	9
Dominance (4 points)	6	7	10	10	65	66
Total	100	102	100	102	100	102

Table 23　Intercorrelations of Circles Test Variables

Dominance Scores (Circles Test)	Relatedness Males	Relatedness Females	Past Dominance Males	Past Dominance Females	Present Dominance Males	Present Dominance Females
Relatedness	. . .	. . .				
Past dominance	−.10	.04	. . .	. . .		
Present dominance	.01	.05	−.32*	−.18	. . .	. . .
Future dominance	−.18	−.38*	−.38*	−.12	−.20*	−.11

*$p < .01$.

234

Table 24 Temporal Relatedness and Future Dominance

| Future Dominance | Temporal Relatedness | | | | |
	Atomistic	Continuous	Integrated-Projected	Total	%
Males [a]					
Absence (0 points)	26 (18)	24 (36)	9 (22)	59	23.6
Secondary (2 points)	25 (17)	10 (15)	8 (19)	43	16.9
Dominance (4 points)	95 (65)	33 (49)	24 (58)	152	59.4
Total	146	67	41	254	
Percentage	57.5	26.4	16.1		100.0
$\chi^2 = 7.99$, $p < .10$ (4 df)					
Females [a]					
Absence (0 points)	14 (21)	11 (41)	3 (33)	28	27.5
Secondary (2 points)	12 (18)	2 (7)	1 (11)	15	14.7
Dominance (4 points)	40 (61)	14 (52)	5 (56)	59	57.8
Total	66	27	9	102	
%	64.7	26.5	8.8		100.0
$\chi^2 = 4.65$, p = n.s. (4 df)					

[a] The number of respondents in a particular category. Percentages are given in parentheses.

235

Table 25 Temporal Relatedness and Present Dominance

	Temporal Relatedness				
Present Dominance	Atomistic	Continuous	Integrated-Projected	Total	%
Males [a]					
Absence (0 points)	58 (40)	32 (48)	14 (34)	104	41.3
Secondary (2 points)	70 (48)	24 (36)	25 (61)	119	46.9
Dominance (4 points)	17 (12)	11 (16)	2 (5)	30	11.8
Total	145	67	41	253	
Percentage	57.5	26.4	16.1		100.0
$\chi^2 = 7.61$, $p < .10$ (4 df)					
Females [a]					
Absence (0 points)	16 (25)	6 (22)	1 (11)	23	23.5
Secondary (2 points)	44 (67)	17 (63)	7 (78)	68	66.7
Dominance (4 points)	6 (8)	4 (15)	1 (11)	11	9.8
Total	66	27	9	102	
%	64.7	26.5	8.8		100.0
$\chi^2 = 2.01$, $p = $ n.s. (4 df)					

[a] The number of respondents in a particular category. Percentages are given in parentheses.

Table 26 Temporal Relatedness and Development

| | Temporal Relatedness | | | | |
Development	Atomistic	Continuous	Integrated-Projected	Total	%
Males [a]					
No	80 (55)	46 (69)	20 (49)	146	57.7
Yes	65 (45)	21 (31)	21 (51)	107	42.3
Total	145	67	41	253	
%	57.3	26.5	16.2		100.0
$\chi^2 = 5.01$, $p < .05$ (2 df)					
Females [a]					
No	26 (39)	13 (48)	4 (44)	43	42.2
Yes	40 (61)	14 (52)	5 (56)	59	57.8
Total	66	27	9	102	
%	64.7	26.5	8.8		100.0
$\chi^2 = 0.62$, $p =$ n.s. (2 df)					

[a] The number of respondents in a particular category. Percentages are given in parentheses.

Table 27 Correlations of Circles Test and Duration Variables for Men ($N = 166$) and Women ($N = 133$)

	Circles Test Variables									
	Relatedness		Past Size		Present Size		Future Size		Development	
Duration Variables	M	F	M	F	M	F	M	F	M	F
Present backward	.02	−.21*	−.06	−.13	.02	.06	.08	−.00	−.05	−.00
Present forward	−.10	−.01	−.01	−.04	.03	.02	.13	−.13	−.01	−.08
Distant past	−.08	−.07	−.08	−.12	−.21**	−.05	.28**	.03	.15	.09
Near past	−.13	.01	.06	−.19*	−.15	.08	.12	−.02	−.03	−.02
Distant future	−.16*	−.15	−.04	−.05	.04	.00	.26**	.02	.15	−.01
Near future	−.27**	−.09	.04	.04	−.07	−.04	.16*	−.14	.02	−.13

*$p \leqslant .05$, two-tailed test.
**$p \leqslant .01$, two-tailed test.

Table 28 Austrian Population Distribution by Sex, Age, and Social Class

| Age Group | Males (N = 101) | | Females (N = 79) | | Total | %[a] |
	Upper Class	Middle Class	Upper Class	Middle Class		
Young (12–15)	13.9[a] (25)	10.0 (18)	9.4 (17)	11.7 (21)	81	45.0
Old (16–18)	13.3 (24)	18.9 (34)	8.9 (16)	13.9 (25)	99	55.0
%	27.2	28.9	18.4	25.5		100.0
Total	49	52	33	46	180	

[a] Percentage of total Austrian sample.

Table 29 Means and Standard Deviations of Lines Test Variables (Austria)

| Variables | Means[a] | | Standard Deviations | | Median Combined | Range Combined |
	Males	Females	Males	Females	Males-Females	Males-Females
Life time	14.98	14.56	4.52	4.67	16.31	0.20–20.00
Present	1.29	1.84	1.84	2.24	0.81	0.10–12.7
Personal past	3.6	3.89	1.87	1.97	3.60	0.10– 9.5
Personal future	9.9	8.85	4.22	4.08	10.00	0.10–17.4
Historical past	2.9	3.00	2.57	2.22	2.30	0.10–13.4
Historical future	3.31	3.40	3.36	3.08	2.25	0.10–16.1

[a] All values refer to centimeters.

Table 30 Intercorrelation Matrix of Lines Test Variables (Austria)

Lines Test Variable	Life Time		Present		Past		Future		Historical Past	
	Males	Females	Males	Females	Males	Females	Males	Females	Males	Females
Life time	...	...								
Present	.12	.19	...	...						
Personal past	.56**	.57**	.24*	.11	...	...				
Personal future	.75**	.80**	-.41**	-.24*	.04	.14	...	...		
Historical past	-.60**	-.84**	-.11	-.09	-.30**	-.56**	-.44**	-.68**	...	...
Historical future	-.86**	-.92**	-.12	-.22*	-.51**	-.47**	-.62**	-.73**	.37**	.56**

$*p \leq .05.$
$**p \leq .01$

240

Table 31   Analysis of Variance of Age, Sex, Social Class, and Life Time Extension

Effect	Sum of Squares	df	Mean Square	F-ratio
Sex	8.69	1	8.69	.43
Age	.3	1	.3	.01
Class	93.1	1	93.1	4.63*
Sex × age	127.37	1	127.37	6.34*
Sex × class	131.74	1	131.74	6.56*
Age × class	2.61	1	2.61	.13
Sex × age × class	.33	1	.33	.02
Error	3454.9	172	20.09	

*$p < .05$.

Table 32   Analysis of Variance of Age, Sex, Social Class, and Present Extension

Effect	Sum of Squares	df	Mean Square	F-ratio
Sex	13.52	1	13.52	3.27
Age	17.73	1	17.73	4.30*
Class	.44	1	.44	.11
Sex × age	1.90	1	1.90	.46
Sex × class	1.57	1	1.57	.38
Age × class	8.05	1	8.05	1.95
Sex × age × class	6.17	1	6.17	1.49
Error	710.70	172	4.13	

*$p < .05$.

Table 33   Analysis of Variance of Age, Sex, Social Class, and Personal-Past Extension

Effect	Sum of Squares	df	Mean Square	F-ratio
Sex	4.70	1	4.70	1.34
Age	17.80	1	17.80	5.07*
Class	6.04	1	6.04	1.72
Sex × age	17.37	1	17.37	4.95*
Sex × class	.62	1	.62	.18
Age × class	11.41	1	11.41	3.25
Sex × age × class	2.01	1	2.01	.57
Error	603.94	172	3.51	

*$p < .05$.

Table 34   Analysis of Variance of Age, Sex, Social Class and Personal-Future Extension

Effect	Sum of Squares	df	Mean Square	F-ratio
Sex	55.90	1	55.90	3.39
Age	4.15	1	4.15	.25
Class	39.50	1	39.50	2.40
Sex × age	34.61	1	34.61	2.10
Sex × class	179.71	1	179.71	10.91*
Age × class	7.55	1	7.55	.46
Age × sex × class	14.33	1	14.33	.87
Error	2833.99	172	16.48	

*$p < .01$.

Table 35   Correlations of Historical Extensions (Lines Test) and Historical Time Values of the Money Game (America)

Money Game Variables	Historical Past		Historical Future	
	Males	Females	Males	Females
Historical past				
1 hour	.27*	.28*		
1 day	.20	.14		
1 year	.28*	.16		
Historical future				
1 hour			−.09	.03
1 day			−.08	.05
1 year			−.10	−.09

*$p < .05$, two-tailed test.

243

Table 36 Correlations of Circles Test with Lines Test Variables (America)

	Circles Test Variables									
	Total Circle Integration		Past Size		Present Size		Future Size		Development	
Lines Test Variables	Males	Females	Males	Females	Males	Females	Males	Females	Males	Females
Life time	-.02	.07	-.04	-.08	.08	-.20*	.11	.08	.07	.33**
Present	.02	-.05	.05	-.13	-.08	.06	.09	-.20*	.11	-.02
Future	-.04	.09	-.10	-.07	-.14	-.04	.09	.07	-.01	.08
Past	.01	.01	.04	-.01	-.04	-.04	-.04	.09	.05	.05
Historical past	.07	-.04	-.04	.07	.04	.27**	-.08	-.07	-.11	-.15
Historical future	-.03	-.09	.07	.05	.14	.13	-.09	-.07	.05	-.06

*$p < .05$, two-tailed test.
**$p < .01$, two-tailed test.

244

Table 37 Range, Percentage, and Frequency Distributions for Life Time Extensions (America)

	Life Time Extensions					
	Historiocentric 0.05–10 cm		Normal 10.5–17.5 cm		Egocentric 18–20 cm	
Typology Metric Range	Males	Females	Males	Females	Males	Females
% distributions	26	25	31	39	43	36
Total	109	24	130	37	180	34

Table 38 Relationship of Life Time Extensions (Lines Test) and Present Dominance (Circles Test) (America)

Present Dominance	Life Time Extensions					
	Historiocentric[a]	Middle[a]	Egocentric[a]	%	χ^2	df
Males						
Absence (0 points)	44 (40)	49 (38)	72 (40)	39.8		
Secondary dominance (2 points)	48 (44)	71 (54)	98 (54)	51.2	9.28*	4
Dominance (4 points)	17 (16)	10 (8)	11 (6)	9.0		
Total	109	130	181			
%	26.0	31.0	43.1			
Females						
Absence (0 points)	5 (21)	11 (30)	7 (21)	24.2		
Secondary dominance (2 points)	13 (54)	24 (65)	25 (73)	65.3	7.97**	4
Dominance (4 points)	6 (25)	2 (5)	2 (6)	10.5		
Total	24	37	34			
%	25.3	38.9	35.8			

[a] The number of respondents in a particular category. Percentages are given in parentheses.

246

Table 39   Means and Standard Deviations of Present Extension, Duration Responses of Present Borders, Present Evaluation and Potency, and Future Preknowledge Fantasy Scores

Variable	Instrument	Males Mean	Males SD	Females Mean	Females SD
Present extension	Lines Test	1.97	3.48	2.58	3.74
Past border	Duration Inventory	2.92	2.30	3.51	2.46
Future border	Duration Inventory	3.19	2.43	3.96	2.44
Relatedness	Circles Test	2.61	3.56	2.12	3.21
Evaluation	Semantic Differential	4.56	1.52	5.04	1.56
Potency	Semantic Differential	4.92	1.04	5.02	1.08
Hour of preknowledge	Money Game	2.37	1.56	1.93	1.32
Day of preknowledge	Money Game	2.72	1.67	2.23	1.40
Year of preknowledge	Money Game	3.14	1.88	2.70	1.77

Table 40 Intercorrelation Matrix of Six Present Variables and Summed Preknowledge Fantasy Score

Time Perception Variables	Present Extension		Past Border		Future Border		Relatedness		Evaluation		Potency	
	Males	Females	Males	Females	Males	Females	Males	Females	Males	Females	Males	Females
Present extension (Lines Test)[a]	...	...	...	...	...	...	...	...	...	...	...	...
Past border (Duration Inventory)	.20*	.13	...	...	...	...	...	...	...	...	...	...
Future border (Duration Inventory)	.24*	.16	.67*	.69*	...	...	...	...	...	...	...	...
Relatedness (Circles Test)	−.11	−.26**	−.08	−.17	−.10	−.04	...	...	...	...	...	...
Evaluation (Semantic Differential)	−.01	−.12	−.01	.08	.00	.18	.10	.06	...	...	...	...
Potency (Semantic Differential)	.07	.08	.12	.10	.06	.21**	−.07	.01	.25*	.37*	...	...
Preknowledge sum (Money Game)	.11	.06	.11	.06	.07	.07	−.08	−.04	−.00	−.32*	.11	−.14

*p < .01 two-tailed.
**p < .05 two-tail.
[a] Description in parentheses refers to instrument.

Table 41 Summary of Relationships between Present Extension (Lines Test), Remaining Present Perceptions, and Personal Future Preknowledge Fantasy Scores

Characteristic of (Dependent) Temporal Variable	Males Present Extension[a]						Females Present Extension[a]					
	Short	Long	%[b]	χ^2	p	df[c]	Short	Long	%[b]	χ^2	p	df[c]
Short past border (seconds, minutes, hours)	168 (73)	73 (50)	64	15.16	.001	1	24 (63)	23 (44)	52	2.47	.12	1
Short future border (seconds, minutes, hours)	166 (72)	66 (45)	61	19.72	.001	1	23 (58)	23 (45)	51	0.987	n.s.	1
Atomicity	117 (51)	102 (70)	59				19 (48)	42 (80)	67			
Continuity	78 (34)	26 (18)	28	14.24	.001	2	17 (44)	5 (9)	23	14.80	.001	2
Integration-projection	34 (15)	17 (12)	13				3 (8)	6 (11)	10			
High evaluation (above mean)	113 (49)	82 (56)	52	1.30	n.s.	1	29 (75)	36 (70)	72	0.044	n.s.	1
High potency (above mean)	99 (43)	86 (59)	49	7.85	.005	1	17 (43)	31 (60)	53	1.46	.23	1
Symmetrical present	156 (68)	73 (50)	61				29 (74)	24 (47)	59			
Present in advance of itself	44 (19)	51 (35)	25	11.62	.003	2	9 (23)	21 (40)	33	6.99	.03	2
Present behind itself	30 (13)	22 (15)	14				1 (3)	7 (13)	8			
High preknowledge (fantasizers)	110 (48)	88 (60)	53	5.06	.02	1	13 (33)	25 (48)	42	1.43	n.s.	1
Total	230	146					39	52				
%	61.2	38.8					42.9	57.1				

[a] The number of respondents in a particular category. Percentages of respondents demonstrating characteristics of both independent and dependent variables are given in parentheses.

[b] Marginal percentage of dependent variable.

[c] With 1 df, Yates chi square values with continuity corrected are presented.

Table 42   VAch Scale

Planning only makes a person unhappy since plans hardly ever work out anyhow.	AGREE	DISAGREE [a]
When a man is born, the success he is going to have is already in the cards, so he might as well accept it and not fight it.	AGREE	DISAGREE
Nowadays, with world conditions as they are, the wise person lives for today and lets tomorrow take care of itself.	AGREE	DISAGREE
Even when teenagers get married, their main loyalty still belongs to their fathers and mothers.	AGREE	DISAGREE
When the time comes for a boy to take a job, he should stay near his parents, even if it means giving up a good job opportunity.	AGREE	DISAGREE
Nothing in life is worth the sacrifice of moving away from your parents.	AGREE	DISAGREE
The best kind of job is one where you are part of an organization where everyone works together, even if you don't get individual credit for the work you do.	AGREE	DISAGREE

[a] Underlining refers to precoded achievement response.

Table 43   Taylor Manifest Anxiety Scale (21-Item Brief Form)

I do not tire quickly.	AGREE	<u>DISAGREE</u> [a]
I believe that I am no more nervous than others.	AGREE	<u>DISAGREE</u>
I work under a great deal of tension.	<u>AGREE</u>	DISAGREE
I worry quite a bit over possible misfortunes.	<u>AGREE</u>	DISAGREE
I feel hungry almost all of the time.	<u>AGREE</u>	DISAGREE
I dream frequently about things that are best kept to myself.	<u>AGREE</u>	DISAGREE
I am easily embarrassed.	<u>AGREE</u>	DISAGREE
I am more sensitive than most other people.	<u>AGREE</u>	DISAGREE
I wish I could be as happy as others seem to be.	<u>AGREE</u>	DISAGREE
I am happy most of the time.	AGREE	<u>DISAGREE</u>
It makes me nervous to have to wait.	<u>AGREE</u>	DISAGREE
Sometimes I become so excited that I find it hard to go to sleep.	<u>AGREE</u>	DISAGREE
I must admit that I have at times been worried beyond reason over something that really did not matter.	<u>AGREE</u>	DISAGREE
I have very few fears compared with my friends.	AGREE	<u>DISAGREE</u>
I have been afraid of things and people that I know could not hurt me.	<u>AGREE</u>	DISAGREE
I find it hard to keep my mind on my task or job.	<u>AGREE</u>	DISAGREE
I certainly feel useless at times.	<u>AGREE</u>	DISAGREE
I am unusually self-conscious.	<u>AGREE</u>	DISAGREE
I am inclined to take things hard.	<u>AGREE</u>	DISAGREE
Life is a strain for me much of the time.	<u>AGREE</u>	DISAGREE
I am entirely self-confident.	AGREE	<u>DISAGREE</u>

[a] Underlining refers to precoded anxiety response.

Table 44 Temporal Anxiety Item Loadings of Temporal Attitude Inventory Factor Analysis

		Factors				
Item Number and Statement	Temporal Anxiety I	Temporal Depreci- ation II	Fantasy Tolerance III	Calculable Time IV	Inconstant Present V	Optimistic Involvement VI	h^2
32. I'm afraid I won't be able to lead a full life.	.65	.04	– .19	– .04	.04	– .09	.47
31. The pace of life is too fast for me.	.61	– .01	– .00	.04	– .03	– .01	.37
19. I'm often so worried about what's going to happen, I forget about right now.	.55	– .00	.03	– .12	– .06	– .18	.35
33. It's hard for me to work on a task when I know there's not enough time to finish.	.53	.12	.00	– .01	– .04	.08	.30
34. I live in the past.	.50	.01	.20	.05	– .04	– .05	.29

39. A minute seems too small an amount of time to be of any use.	.45	.28	.06	.19	− .21	− .02	.35
23. I dislike change.	.45	.33	.03	.11	.24	.16	.40
18. I live in the future.	.45	− .11	.08	.01	.01	.34	.34
20. More than knowing what's happening in the world, I like to know *why* things happen.	.41	.19	.07	.17	.24	− .06	.30
1. I sometimes feel I'm racing time and losing.	.40	.06	.02	− .26	−.15	− .26	.32
11. Often I think how nice it would be to stop time.	.39	.07	.33	− .28	.15	.01	.37
15. I live in the present.	− .39	.14	− .17	− .33	.07	.25	.37
Average	48.2	11.3	9.8	13.4	10.7	12.6	35.2
% common variance	30.4	15.1	15.4	13.3	12.5	12.4	

Table 45 Means and Standard Deviations of Variables

Variables	Males		Females	
	Mean	SD	Mean	SD
Independent				
Intelligence	113.4	11.72	109.6	10.22
VAch	5.62	1.11	5.41	1.03
TMAS	8.73	4.14	9.39	4.09
Temporal				
TAS	15.15	6.29	14.48	6.19
Experiential time orientation	2.34	0.654	2.38	.624
Future potency	5.12	1.03	5.11	1.15
Future evaluation	5.18	1.36	5.47	1.55
Past recovery [a]	6.84	1.62	6.38	1.58
Future preknowledge [a]	6.74	1.70	6.29	1.50

[a] Values now summed for the three allotments (cf. Table 41).

Table 46 Intercorrelations of VAch, TMAS, Intelligence and the Temporal Variables

Temporal Variables	VAch		TMAS		Intelligence	
	Males	Females	Males	Females	Males	Females
Experiential time orientation	.07	.19	−.04	.07	.06	.15
Total relatedness	.25**	.26*	−.26*	.11	.21**	.20*
Temporal anxiety	−.28**	−.28*	.55**	.48**	−.01	−.01
Present past duration	−.02	.18	−.02	−.07	−.10	−.09
Present future duration	−.02	.20*	.01	.21*	−.13	−.13
Future potency	−.08	−.05	.07	.05	−.00	−.15
Future evaluation	.06	.16	−.04	−.06	.10	.08
Past recovery	−.10	−.08	.26*	.12	−.11	−.09
Future preknowledge	−.07	.02	.01	.16	−.02	−.08

*$p < .05$.
**$p < .01$.

Table 47 Intercorrelations of VAch, TMAS, and Intelligence

Variables	IQ		VAch		TMAS	
	Males	Females	Males	Females	Males	Females
Intelligence	. . .	. . .				
VAch	.24**	.23*	. . .	. . .		
Anxiety (TMAS)	−.04	−.07	−.14	−.22*	. . .	. . .

*$p < .05$.
**$p < .01$.

255

Table 48 Relationship between VAch and Selected Temporal Variables Controlled by Intelligence

Characteristic of Temporal Variable	Low Intelligence				High Intelligence				
	Low VAch[a]	High VAch[a]	%[b]	χ²	Low VAch[a]	High VAch[a]	%[b]	χ²	df[c]
Males									
Semantic differential									
High future potency (above mean)	45 (63)	31 (42)	52.3	4.58*	31 (58)	62 (47)	50.6	1.21	1
High future evaluation (above mean)	37 (52)	26 (35)	43.0	3.01**	29 (54)	69 (52)	52.4	0.014	1
Experiential time orientation									
Future avoidant	18 (25)	16 (21)	22.9		22 (41)	44 (33)	34.9		
Middle	28 (39)	26 (35)	37.1	0.816	20 (37)	34 (26)	29.0	6.43*	2
Future orientation	25 (36)	33 (44)	40.0		12 (22)	54 (41)	36.0		
Total	71	75			54	132			
%	48.4	51.6			28.9	71.1			

256

Females

Present duration									
Long future border (above mean)	12 (47)	21 (63)	54.4	0.935	4 (20)	13 (63)	44.1	4.70*	1
Long past border (above mean)	10 (39)	19 (59)	48.3	1.69	5 (27)	12 (58)	44.1	2.17	1
Temporal relatedness									
Atomicity	20 (78)	22 (68)	73.3	4.91	9 (50)	11 (52)	51.3	4.99**	2
Continuity	6 (22)	6 (18)	20.0		9 (50)	5 (26)	35.9		
Integration-projected	0 (0)	5 (14)	6.7		0 (0)	5 (22)	12.5		
Total	26	33			18	21			
%	44.4	55.6			45.2	54.8			

[a] The number of respondents in a particular category. Percentages are given in parentheses.
[b] Marginal percentage of temporal variable.
[c] With 1 df, Yates chi square values with continuity corrected are presented.
*p < .05.
**p < .10.

Table 49 Relationship Between TMAS and Selected Temporal Variables Controlled by Intelligence

Characteristic of Temporal Variable	Low Intelligence Anxiety					High Intelligence Anxiety					
	Low[a]	Medium[a]	High[a]	%	χ^2	Low[a]	Medium[a]	High[a]	%	χ^2	df[b]
Semantic differential											
High future potency (above mean)	25 (49)	22 (46)	30 (65)	53.0	3.12	30 (47)	30 (54)	34 (52)	51.0	.556	2
High future evaluation (above mean)	21 (41)	23 (49)	16 (35)	41.7	1.39	27 (41)	36 (64)	40 (60)	54.4	5.68**	2
Temporal relatedness											
Atomicity	20 (39)	34 (73)	34 (73)	60.9		28 (43)	28 (50)	36 (54)	49.0		
Continuity	23 (46)	11 (24)	9 (19)	30.4	13.68**	24 (37)	13 (23)	16 (25)	28.6	3.39	
Integrated-projected	8 (15)	2 (3)	4 (8)	8.7		13 (20)	15 (27)	14 (21)	22.4		4
Temporal anxiety											
High TMAS (above mean)	18 (36)	27 (58)	35 (75)	55.1	11.09**	5 (23)	30 (54)	53 (80)	51.7	26.84**	2
Temporal fantasy											
Recovery fantasizers	26 (51)	22 (47)	27 (57)	51.8	.672	24 (37)	27 (48)	43 (65)	50.3	8.32*	2
Preknowledge fantasizers	28 (55)	25 (53)	28 (59)	55.8	.344	25 (39)	35 (63)	36 (54)	51.4	5.39*	2
Total	51	47	47			65	56	66			
%	35.7	32.2	32.2			34.7	29.9	35.4			

Females

Temporal anxiety							
High TAS	5 (22)	14 (55)	39.5	2.99	5 (23)	23 (84)	62.9	6.24**	1
Total	23	25			23	27		
%	47.4	52.6			45.7	54.3		

[a] The number of respondents in a particular category. Percentages are given in parentheses.
[b] With 1 df, Yates chi-square values with continuity corrected are presented.

*p < .05.
**p < .01.
***p < .10.

INDEX

Achievement;
  anxiety negatively correlated with, 159
  intelligence related to, 158
  perception of time affected by, 149–64
Achievement Value Inventory (VAch), 154,
  155
Age;
  denial of past in, 181–82
  as variable in Lines Test, 111–16, 119, 120,
  123
Allport, Gordon, on future orientation,
  67–69, 79
Analysis of variance, of Lines Test, 113–16
Anxiety;
  achievement negatively correlated with,
  159
  Freud on, 154, 161, 165n.
  Heidegger on, 167n.
  intelligence not related to, 159
  perception of time affected by, 149–64
Appolonian epoch, 178–79
Army General Classification Test, 158, 166n.
Asymmetric time, 41
Atomicity, *see* Temporal atomicity
Augustine, Saint, vii, 22
  on present, 20, 86, 92, 132
Austrian students, as sample group, 112–16,
  118, 120, 121, 143

Bakan, David, 183
Banfield, Edward, 127n.
Barrett, William, 161
Becket, Thomas, 180
Beckett, Samuel, 40
Benford, F., 96
Bergson, Henri, 2, 30, 147n., 148n.
  on duration of time, 38, 48n., 102–3, 125n.
  Kummel on, 18n.
  on present, 142, 190n.
  Schutz on, 170
  spatial conception of time of, 177
  on time as continuous flow, 10–12

on time zones, 92, 99n.
on true self, 21–23
Binswanger, Ludwig, 23, 66, 162
  on schizophrenia, 151–52
Bonier, R.T., 104
Boss, Medard, 32n.
  on fantasies, 64n.–65n.
Boundaries between past, present and
  future, 137–41
  achievers' perception of, 153, 158
  anxious people's perception of, 153, 160
Broad, 126n.
Brumbaugh, Robert, 69
Bruner, Jerome, 18n.

Calvinism, 61
Cartwright, L., 163
Cassirer, Ernst, 17n., 18n., 99, 148n.
Certainty-Uncertainty Scale, 73–74
Childhood, 175, 191n.
Circles Test, 2, 87–99
  achievement value and, 151–54, 158, 160
  anxious people on, 153, 158, 160, 161
  atomicity on, 138, 139, 142, 144–45, 159,
  184
  Duration Inventory and, 89, 95, 122
  Lines Test and, 109, 119, 122–24
  present in, 123, 124, 133, 137–42, 144–45,
  174
  sex differences in, 185–86
  temporal relatedness in, 87–91, 158
  time zones in, 123, 138, 144–45, 174, 175,
  193n.
Class, *see* Middle class; Social class; *and*
  Upper class
Clay, E. R., 146n.
Cohen, John, 34, 103, 104
Collingwood, Robin George, 7
Competence, White's concept of, 152
Continuity, *see* Temporal continuity
Cournot, Antoine Augustin, 58

Dascin (existentialist concept), 130n.
Death:
    future-oriented people's preoccupation
        with, 61
    philosophers on, 57–58
Decoupling, 128n.
Descartes, René, 11
de Tocqueville, Alex, 128n.
de Unamuno, Miguel, 58
Dewey, John, 78, 171, 181
Dilthey, Wilhelm, 123
Discontinuity, see Temporal discontinuity
Distant Cosmic Commitment Scale, 74
Dogmatism, 73, 83n., 104
Doob, Leonard, 52, 65n.
Dooley, Lucille, 179, 188
Dummett, Michael, 173, 190n.
Duration:
    Bergson on, 38, 48n., 102–3, 125n.
    experiments in, 6–7, 103–4, 107, 108
    Lines Test of, 104–24
    of present, 9–10, 131, 137, 141
Duration Inventory, 2, 38–47
    achievers on, 153, 158
    anxious people on, 153, 160
    atomism in, 176–77
    Circles Test and, 89, 95, 122
    future commitment and, 77
    Lines Test and, 139
    present in, 40, 132, 137–42, 172, 173
    time zones in, 163, 175

Education, as controlled variable, 3–4
Ego, 179, 188
Egocentric perception:
    future in, 177–78
    on Lines Test, 108–11, 114, 116, 117,
        120–24
Einstein, Albert, 187
Eissler, Kurt, 178–79, 192n.
Eliade, Marcea, 65n., 102, 181
Emerson, Ralph Waldo, vii
Epley, David, 150
Erikson, Erik, 178, 188
    on identity, 97, 187
    on present, 146n.
    on sex role of women, 129n.
Eson, Morris, 15, 35, 46
Exchange, in Money Game, 63–64
Existentialists, 125n.
    Dascin concept of, 130n.

Expectations, Parsons on, 78
Experience, different interpretations of,
    169–71
Experiental Inventory, 1, 25–31, 36
    achievers on, 153, 159
    anxious people on, 153, 160
    different interpretations on, 169–72
    duration of time and, 43–46
    future orientation on, 44, 69–71, 80, 143,
        149, 159, 171–72, 177
    Lines Test and, 109
    time zones in, 25–29, 169, 170
Experiments:
    Achievement Value Inventory, 154, 155
    Circles Test, 87–99
    on duration, 6–7, 103–4, 107, 108
    Duration Inventory, 38–47
    Experiental Inventory, 25–31, 36
    Future Commitment Scale, 73–77
    Lines Test, 104–24
    Money Game, 51–64
    Semantic Differential, 71–73, 75–76
    Taylor Manifest Anxiety Scale, 154, 155,
        158, 159
    Temporal Anxiety Scale, 154–59
    See also individual scores and tests
Expressive interaction, 186
    future orientation and, 78–81, 90
Extended present, 142–45, 174
Extension, 146n.

Factor analysis:
    of Future Commitment Scale, 74, 157
    of Temporal Anxiety Scale, 156–57
Fantasizers (in Money Game), 60, 71, 138,
    159
Fantasy, 2, 50, 179–80, 184
    in anxious persons, 152, 158
    Boss on, 64n.–65n.
    Meerloo on, 53, 55
    in Money Game, 51–64, 71, 138, 159
Farnham-Diggory, Sylvia, 104, 107
Faustian epoch, 178–79
Femininity:
    expressive interaction and, 78
    sex role of, 111, 183
    see also Sex; Sex roles; and Women
Fingarette, Herbert, 51
Fourgerousse, Carl Edward, 161
Fraisse, Paul, 20–21, 108
    on present, 132, 133, 146n.

Frank, Jerome, 123–24
Frank, Lawrence, 30
Freud, Sigmund, 162
  on anxiety, 154, 161, 165n.
Fundamental self, Bergson on, 102–3
Future:
  achievement value and perception of,
    150–54, 158
  boundary between present and, 137–41,
    153, 158
  as continuation of present, 133–34
  in egocentric perception, 177–78
  Heidegger on, 57–58
  Kelly on, 23–24, 26, 30, 66, 67
  Kluckhohn on, 93
  in linear conception of time, 68
  in Money Game, 55, 57–58, 60–62, 137–39,
    141, 142, 174
  to philosophers, 13–14, 23–24, 66–70
  in religious belief, 58, 61, 129n.
  schizophrenia and, 151–52
  in spatial conception of time, 68
  in studies, 35–36
  see also Historical future; Personal future
Future avoidance, 71
  on Experiental Inventory, 159
  on Future Commitment Scale, 74–75
Future Commitment Scale, 2, 73–77, 184
  factor analysis of, 74, 157
Future dominance, 23, 96
Future-dominant development, 93–94
Future orientation, 66–81, 143–144, 163
  achievement need and, 149
  in anxious people, 162
  death preoccupation and, 61
  egocentric perception and, 124
  on Experiental Inventory, 44, 69–71, 80,
    143, 149, 159, 171–72, 177
  expressive interaction and, 81, 90
  on Future Commitment scale, 74–75
  instrumental interaction and, 77–78,
    80–81, 90, 144, 186
  Semantic Differential and, 71–73, 75–76
  sex roles and, 77–80

General Classification Intelligence Test,
  158, 166n.
Gestaltist reasoning, 176
Gestalt theory, 146n., 191n.
Gioscia, Victor, 176
Greenfield, Norman, 35

Gurvitch, Georges, 41, 101n., 133
  Circles Test and, 100n.
Guyau, Marie Jean, 146n.–147n.

Hall, Edward, 93–94, 96
Harton, John, 15, 19n.
Harvitz, Howard (fictional character), 157
Hatcher, Anne, 79
Hegel, G.W.F., 58
Heidegger, Martin, 22, 125n., 167n.
  on anxiety, 167n.
  authentic man concept of, 161
  on existence, 105
  on future, 57–58
  on present, 90, 132, 190n.
  on remembering, 86
  on responsibility, 162
  on understanding, 31n.–32n.
Henry, William, 51, 150
Hensel, C.E.M., 104
Hermann, Robert S., 73
Heterogeneous duration, 102
Historical future:
  in Lines Test, 106–7, 116, 121
  in Money Game, 57–58
  see also Future
Historical past:
  in Lines Test, 106–7, 116, 121
  in Money Game, 55–57
  as sacred time, 180–81
  see also Past
Historiocentric perception:
  on Lines Test, 108, 110, 111, 114, 116–18,
    122–24
  present in, 178
History:
  in perception of time, 7
  see also Historical past; Past
Hobbes, Thomas, 21
Homogeneous duration, 102
Husserl, Edmund, 14, 18n., 22
  on death, 57

Id, 179
Identity, Erikson on, 97, 187
Immediate Social Commitment Scale, 74
Instantaneous present, 114, 142, 143, 174
Instrumental interaction, future orientation
  and, 77–78, 80–81, 90, 144, 186
Integration, see Temporal integration
Intelligence tests:

achievement and anxiety measures and, 157–60, 166n.
future orientation and differences in, 79–80
not available for Vienna students, 120
perception of time and, 159–60
temporal continuity and, 97
Intention for future, 184

James, William, 165n.
on present, 133, 191n.
saddlebacks of consciousness concept of, 135
Jung, Carl, 172–73, 192n.

Kafka, Franz, 90
Kafka, John, 15
Kant, Immanuel, 11–12
Kelly, George, on future, 23–24, 26, 30, 66, 67
Kiekegaard, Sören, 102, 165n.
Klineberg, Stephen, 66, 190n.
Kluckhohn, Florence, 24, 30, 67
on future, 93
on time orientations, 97
Kracauer, Siegfried, 68, 125n.
Kummel, Fredrich, 8, 12, 19n., 162
on Bergson, 18n.
on duration, 126n.
spatial conception of time and, 14–15, 85–86, 90

Langer, Jonas, 16
Lewin, Kurt, 147n.
on adolescence, 107–8, 115
on intelligence and symbolic thought, 84n.
on interference, 98–99
Life time, in Lines Test, 113–14, 117, 119–23, 134
Lifton, Robert, 180, 190n.
Linear conception of time, 6–17, 50, 102–25
future in, 68
in women, 81, 85
Lines Test, 104–24
Circles Test and, 109, 119, 122–24
Duration Inventory and, 139
present in, 132, 134–35, 137–41, 144

McClelland, David, 86, 149, 150, 161
Macroexperiencing, 135
Magian epoch, 192n.

Malamud, Bernard, 157
Mann, Thomas, 148n.
Mannheim, Karl, 56
Market orientation in time perception, 182–83
Marriage, 185–86, 193n.
Masculinity:
instrumental interaction and, 78
of Navy women, 4
sex role of, 111, 183
see also Men; Sex; and Sex roles
Maslow, Abraham, 80
May, Rollo, 23, 66
Mead, George Herbert, 68, 184
Means and ends, 78
Mediate Global Commitment Scale, 74
Meerloo, Joost A.M., 32n.
on fantasy, 53, 55
Melges, Frederick Towne, 161
Memory, 12–13
Men:
anxiety and achievement in, 158–60, 162
on Circles Test, 139, 142
in duration experiments, 108
on Duration Inventory, 45–47, 138, 141, 142, 181
on Experiental Inventory, 71, 162
on Future Commitment Scale, 74, 75
instantaneous present in, 142–44
intelligence and future orientation in, 79
on Lines Test, 109, 111, 113–19, 122–23, 138, 141
in Money Game, 54, 55, 58–61, 63–64, 139, 142, 158
perception of time, contrast with women's, 169–89
on Semantic Differential, 73, 139, 142, 162
sex roles and future orientation in, 77, 79–81
spatial conception of time used by, 85
temporal relatedness and temporal dominance in, 91–92, 95–97
see also Sex; Sex roles
Merlan, Philip, 105, 162
Merleau-Ponty, Maurice, 57
Methodology:
in Achievement Value Inventory, 155
in Circles Test, 87–90
in duration experiments, 103–4
in Duration Inventory, 38–39
in Experiental Inventory, 25–26

in Future Commitment Scale, 73–74
in Lines Test, 104–7
in Money Game, 52
population used, 3–5
in Semantic Differential, 72
in Taylor Manifest Anxiety Scale, 155
in Temporal Anxiety Scale, 155–57
see also individual scores and tests
Microexperiencing, 135
Middle class:
in Lines Test, 110–12, 114–19
see also Social class
Minkowski, Eugene, 34, 101n.
Moments, in linear conception of time, 102
Money Game, 2, 51–64
achievers in, 152, 154
anxious people in, 154, 158, 160
future in, 137–39, 141, 142, 174
future orientation in, 71
Lines Test and, 121
sex differences in, 183
Moore, Wilbert, 174
Mowrer, O. Hobart, 86
Muller, John, 101n.

Navy personnel:
on Lines Test, compared with Vienna
students, 120–21, 123
as sample group, 3–5, 143
Nietzche, Friedrich, 102

Objective perception of time, 7–10, 14, 50
Occupation:
as controlled variable, 3
perception of time affected by, 143
Orientation ascription, 111
Osgood, Charles, 71

Parsons, Talcott, 193n.
on instrumental interaction, 77–78, 80–81,
90, 144, 186–87
Pascal, Blaise, 124
Past:
boundary between present and, 137, 139,
141
in Money Game, 52–57
see also Historical past; Personal past
Past dominance, 22
Past-dominant development, 93
Past orientation, 44, 171, 180
of anxious people, 153

Personal Commitment Scale, 74
Personal future, in Lines Test, 106–7, 115,
117, 118, 120, 121
Personal past, in Lines Test, 106–7, 115, 120,
121
Phenomenology, 22–23
sense of time in, 105, 163
Population samples:
Navy personnel, 3–5, 112
Vienna students, 112–16, 118, 120, 121,
143
Potency of future, in achievers and anxious
persons, 151, 153–54
Prediction of future, 184
on Future Commitment Scale, 76–77
instrumental interaction and, 78
Present:
achievement value and perception of, 150
Augustine on, 20, 86, 92, 132
Bergson on, 142, 190n.
boundary between future and, 137–41,
153, 158
boundary between past and, 137, 139, 141
in Circles Test, 123, 124, 133, 137–42,
144–45, 174
defined, 135, 146n.
duration of, 9–10, 131, 137, 141
in Duration Inventory, 40, 132, 137–42,
172, 173
Erikson on, 146n.
Fraisse on, 132, 133, 146n.
Heidegger on, 90, 132, 190n.
in historiocentric perception, 178
insignificance of, 92
James on, 133, 191n.
in Lines Test, 105–7, 109, 114–15, 118, 124,
132, 134–35, 137–41, 144
perceptions of, 131–45
in refracted time, 62
in Semantic Differential, 137–42, 149, 174
sensation of, 36–38
Whitehead on, 131–32
Whitrow on, 123, 145, 146n.
Present dominance, 22, 122, 123, 179
Present orientation, 141, 171–74
Bergson on, 190n.
in Experiental Inventory, 44
Priestly, J.B., vii, 21
Procedures, *see* Methodology
Projection, *see* Temporal projection
Proust, Marcel, 22

Psychoanalysis, 7
Psychophysical estimation studies, 6–7
Public time, 174

Rapaport, David, 174
Realists (in Money Game), 60, 71, 138, 159
Redfield, Robert, 194n.
Reflective awareness, 174
Refracted time, 62
Religious belief, future in, 58, 61, 129n.
Remembering, Heidegger on, 86
Responsibility:
    Heidegger's concept of, 162
    sense of, 173–75
Results:
    of Achievement Value Inventory, 158
    of Circles Test, 90–99, 139–42, 144, 159,
        160, 175, 177
    of Duration Inventory, 39–42, 139–42, 159,
        160, 172, 175, 181
    of Experiental Inventory, 27–29, 159, 160,
        169–70, 177
    of Future Commitment Scale, 74–77, 184
    of General Classification Intelligence
        Test, 158
    of Lines Test, 112–23, 138–41, 177
    of Money Game, 54–58, 139, 141, 142, 160,
        183
    on perception of present, 138–42
    of Semantic Differential, 72–73, 139–42,
        159, 160
    of Taylor Manifest Anxiety Scale, 158–60
    of Temporal Anxiety Scale, 158, 159
    see also individual scores and tests
Rhythms, 6
Ricks, David, 150
Riesman, David, 124
Roberts, Alan H., 73
Rokeach, Milton, 70
Russell, Bertrand, 18n.

Sacred time, 180–81
Sample groups:
    Navy personnel, 3–5, 112, 143
    Vienna students, 112–16, 118, 120, 121,
        143
Schachtel, Ernest, 24
Schizophrenia, 151–52
Schultz, John, 119
Schutz, Alfred, 172
    on Bergson, 170

Secondary dominance, 123
Self, Bergson on, 21–23, 102–3
Self-actualization, 80
Semantic Differential Scale:
    achievers on, 154, 159
    anxious people on, 153, 160
    future orientation and, 71–73, 75–76
    present in, 137–42, 149, 174
Sex:
    perception of time, contrast between,
        169–89
    as variable in Lines Test, 108, 111–14, 119,
        120
    see also Femininity; Masculinity; Men;
        and Women
Sex roles, 108, 162
    future orientation and, 77–80
    Lines Test differences and, 109, 111, 118
    perception of time and, 182–86
    see also Femininity; Masculinity; Men;
        and Women
Sexual development, 182–83, 193n.
Smart, J.J.C., 8
Smith, John E., 144
Social class:
    as controlled variable, 3
    in Lines Test, 110–20
    see also Middle class; Upper class
Socialization, 119–20
Social mobility, 116, 117
Social role differences, see Sex roles
Social upgrading, 117
Spatial conception of time, 6–17, 50, 85–99
    future in, 68
    in men, 81, 85
Spengler, Oswald, 178, 192n.
Spinoza, Baruch, 11
Strodtbeck, Fred L., 24, 30
    on achievement, 149–50, 154
    Achievement Value Inventory of, 155
Subjective perception of time, 7–10, 14, 50
Sylvester, J., 104

Taylor Manifest Anxiety Scale (TMAS), 154,
    155, 158–60
Temporal anxiety, 179
Temporal Anxiety Scale (TAS), 154–59
Temporal atomicity, 179
    of anxious people, 153, 160, 161
    on Circles Test, 89–91, 93, 97, 138, 139,
        142, 144–45, 159, 184

discontinuity and, 175–77
Temporal continuity, 89, 138, 139, 159,
    175–77
  intelligence and, 97
Temporal development, 92–94, 177
Temporal discontinuity, 175–77
Temporal dominance:
  in Circles Test, 87, 90, 91
Temporal horizon, 8, 175, 177
Temporal integration, 89–90, 98, 138, 139,
    159
  achievement value and, 151
Temporal projection, 89–90, 98, 138, 139, 159
  achievement value and, 151
Temporal relatedness:
  in Circles Test, 87–91, 158
  intelligence and, 97, 158
Tenses (grammer), 63
Thistlethwaite, 110, 117
Tillich, Paul, 105, 125n.
Time:
  achievement and anxiety affecting
    perceptions of, 149–64
  linear and spatial conceptions of, 6–17
  orientations in, 20–31
  as variable, 6–7
Time binding, 86–87
Time lines, 103–4. See also Lines Test
Time zones:
  achievement, anxiety and perception of,
    152
  Bergson on, 92, 99n.
  in Circles Test, 91, 94, 97, 123, 138,
    144–45, 174, 175, 193n.
  in Duration Inventory, 163, 175
  in Experiental Inventory, 25–29, 169, 170
  in Lines Test, 107–9, 113, 118, 123, 163
  perceived duration of, 34–47
  in spatial conception of time, 86–87
Tiryakian, Edward, 56, 98
Tompkins, Sylvan, 147n.
Tradition directedness, 124
True self, Bergson on, 21–23

Updike, John, 126n.
Upper class:
  in Lines Test, 110–12, 114–19
  see also Social Class

Van der Leeuw, G., 11, 36
Variables:
  age, in Lines Test, 111–16, 119–20, 123
  in Circles Test, 87
  controlled, 3–5, 112
  in duration experiments, 104, 111
  sex, in Lines Test, 108, 111–14, 119, 120
  social class, in Lines Test, 110–20
  time as, 6–7
Vienna students, as sample group, 112–16,
    118, 120, 121, 143

Wechsler, David, 79
Werner, Heinz, 16
Wertheimer, Max, 146n.
White, Robert, 98, 152
Whitehead, Alfred North, 22, 29, 30
  on present, 131–32
Whitrow, G. J., 11, 12
  on present, 123, 145, 146n.
Wishes, in fantasies, 53
Witkin, Herman A., 84n.
Wittgenstein, Ludwig, 41, 42
Wohlford, Paul, 147n., 150, 174
Women:
  anxiety and achievement in, 158–60, 162
  on Circles Test, 139–42, 160, 185–86
  in duration experiments, 108
  on Duration Inventory, 44–45, 138, 140,
    142, 158–60, 162, 181
  on Experiental Inventory, 71, 143
  extended present in, 142–43
  on Future Commitment Scale, 74, 75
  on Future Commitment Scale, Duration
    Inventory correlation, 77
  intelligence and future orientation in, 79
  linear conception of time used by, 81, 85
  on Lines Test, 109, 111, 113–19, 122–23,
    138, 140–41
  in Money Game, 54, 55, 58–61, 63–64, 139
  in Navy sample, 4, 143
  perception of time, contrast with men's,
    169–89
  present dominance in, 122
  on Semantic Differential, 72–73, 139, 140
  sex roles and future orientation in, 77–81
  temporal relatedness and temporal
    dominance in, 91–95, 97
  see also Sex; Sex roles